1992

PLANT KAIROMONES IN INSECT ECOLOGY AND CONTROL

CONTEMPORARY TOPICS IN ENTOMOLOGY

Series Editors

Thomas A. Miller
University of California, Riverside
U.S.A.

Helmut F. van Emden
University of Reading
United Kingdom

Plant Kairomones in Insect Ecology and Control

ROBERT L. METCALF
AND
ESTHER R. METCALF

CHAPMAN AND HALL
NEW YORK AND LONDON

First published in 1992 by
Chapman and Hall
an imprint of
Routledge, Chapman & Hall, Inc.
29 West 35 Street
New York, NY 10001-2291

Published in Great Britain by
Chapman and Hall
2-6 Boundary Row
London SE1 8HN

Printed in the United States of America

Library of Congress Cataloguing in Publication Data
Metcalf, Robert Lee, 1916–
Plant kairomones in insect ecology and control / Robert L. Metcalf, Esther R. Metcalf.
p. cm. – (Contemporary topics in entomology)
Includes bibliographical references and index.
ISBN 0-412-01991-4
1. Plant kairomones. 2. Insect baits and repellents. 3. Insect-plant relationships. 4. Insects—Ecology. 5. Chemical ecology. I. Metcalf, Esther. II. Title. III. Series.
SB933.5.M47 1992 91-18099
632′.7—dc20 CIP

British Library Cataloguing in Publication Data also available

IN MEMORIAM

Before the present volume could be printed, Esther Rutherford Metcalf passed away after a lengthy illness. Dr. Metcalf received her Ph.D. degree in nutrition and biochemistry from Cornell University, and for some 40 years worked with her husband, Robert L. Metcalf, on projects spanning the length and breadth of the field of entomology. Their three children earned advanced degrees in education, entomology, medicine, and veterinary medicine. Esther Metcalf also nurtured some 80 graduate students as a second mother or perhaps more appropriately as their mother professor.

Those who were blessed with this privilege will always remember her as gracious, caring and loyal. Such a foundation of support lives in perpetuity by the standard it sets.

The Metcalfs were equally at home in Hawaii working on *Ceratitis capitata* as they were in Illinois where they pioneered studies on *Diabrotica* feeding attractants. These, particularly the latter, are documented in the present book that has come to represent a turning point with the passing of the matriarch of one of the outstanding scientific families in the United States.

CONTENTS

PREFACE

This monograph is the product of more than 40 years of research in the wonderful world of chemical ecology. We were captivated in 1949 when assigned to Hawaii to study the tephritid fruit flies, upon observing the extraordinary response of the oriental fruit fly to the plant kairomone methyl eugenol. We vividly recall walking into our bedroom where a handkerchief with a trace of methyl eugenol was left, to find a screened window several meters away literally black with bemused *Dacus dorsalis* males. Throughout the years our explorations of the Dacinae have become a scientific hobby encompassing 200,000 collective miles of travel to the exotic habitats of the tropical fruit flies in Hawaii, Australia, and Taiwan.

The importance of kairomone lures for monitoring and controlling the tropical fruit flies suggested that similar endeavors with other insect taxa could be equally rewarding. We were fortunate to find here at home in the American Middle West, equally fascinating problems in the chemical ecology of the Diabroticite rootworm beetles that have continued to challenge us for two decades.

From the viewpoint of chemical ecology, the Dacinae and the Diabroticites have much in common. Both groups contain large numbers of rapidly evolving species which are tightly knit together in relationships with plants that are cemented by a variety of secondary plant compounds. In both groups, the great majority of species are obscure and retiring but a few species have emerged to rank among the world's most successful insect pests. For both the Dacinae and the Diabroticites chemical cues provided by simple phenylpropanoid plant kairomones are determinative to ecological behavior. In depth study of the intricacies of these relationships of plants-chemicals-insects inevitably focus on long term evolutionary strategies in a relatively unexplored realm of "chemical evolution".

It is our hope that this monograph will interest other investigators, in both basic and applied studies, of the seminal roles that plant kairomones play in insect behavior, ecology, and evolution. The fascination of trying to understand the evolutionary and ecological aspects speak for themselves. As for the applied, we can do no better than recall Rachael Carson's challenge of 30 years ago to entomologists to devise "new, imaginative, and creative approaches to the problem of sharing our earth with other creatures".

We remain indebted to many colleagues whose names are found in the bibliography for valuable contributions to these research projects, and we are grateful to the U.S. National Science Foundation, the U.S. Department of Agriculture Competitive Research Grants Office,and the California Department of Food and Agriculture for generous research support.

Robert L. Metcalf
Esther R. Metcalf

1

CHEMICAL ECOLOGY OF PLANT KAIROMONES

I. INTRODUCTION

Chemical ecology, defined by its International Society, "is the study of structure, function, and biosynthesis of natural products; their importance at all levels of ecological organization; their evolutionary origin, and their application to social needs". This definition as it relates to the importance of plant secondary compounds as allelochemicals in stimulating and regulating the behavior and ecological interactions of insect herbivores, effectively outlines the subject matter of this monograph. For its preparation, we have drawn upon our previous reviews of insect chemical ecology (Metcalf 1985, 1986, 1990, and Metcalf & Lampman 1989a).

Estimates suggest that there are at least 100,000 chemical compounds produced during the growth and development of the more than 200,000 species of flowering plants. The vast majority of these secondary plant compounds are not essential for the normal physiology of plant growth and reproduction, but rather are the seemingly capricious outpourings of nature's chemical factories. The structures of more than 6000 alkaloids, 3000 terpenoids, several thousands of phenylpropanoids, 1000 flavanoids, 500 quinones, 650 polyacetylenes, and 400 amino acids have already been elucidated. It is only when we carefully examine the evolutionary processes that have brought about the diversification and speciation of plants that this enormous array of organic chemicals takes on orderly and purposeful significance.

The external plant environment is pervaded by these plant-produced biochemicals that ooze from leaves, blossoms, and fruits. While humans are well aware of some of them that define the colors, odors, and tastes that characterize our interactions with plants, there are countless others, less conspicuous in either quantity or quality, that dominate the lives of the 500,000 or so species of insects that have evolved together with the flowering plants. Many of these allelochemicals generate olfactory or gustatory stimuli that convey specific behavioral messages to species involved in ecological interrelations of food webs. From an arthropogenic view, these semiochemicals are perceived by insect sensory receptors as attractants, repellents, arrestants, and phagostimulants. From the viewpoint of the processes of coevolution between plants and insects (Ehrlich & Raven 1964), these allelochemicals are classified as allomones if they convey adaptive advantages to the plant producer, and as kairomones if they convey adaptive advantages to the insect receiver (Kogan 1982).

In an ecological sense, plant odorants dominate the atmospheric chemical environment, pervading terrestrial communities where hundreds of plant species, each with its own characteristic chemical odor spectrum, produce what Wilson (1970) has described as an enormously complex and shifting maze of overlapping active odor spaces, from which thousands of associated insect species must select the few critical signals that stimulate behavioral patterns leading to preferred oviposition sites, to satisfactory food supplies, and to aggregation with receptive mates, or to shelter.

The semiochemical environment of our cultivars must be equally decisive in determining the plant-insect interrelationships of agroecosystems. Here, the number of active spaces generated by plant monoculture is much simpler qualitatively, but much more intense quantitatively. Nevertheless, modern gas chromatography-mass spectrometry (GC-MS) studies of individual plant species have revealed that a spectrum of perhaps 30 to 100 volatiles characterizes the aroma of our cultivars (Metcalf 1987). Fraenkel (1969) reviewed much of the early history of our understanding of the reasons for the existence of plant allelochemicals and suggested criteria for their positive identification as 1) isolation and identification of the chemical, 2) initiation of the allomonal or kairomonal response when the chemical is applied to a neutral surface, and 3) demonstration of a qualitative relationship between the concentration of the allelochemical and the insect behavioral response. The same chemical compound may act as an allomone protecting the plant against attack by an herbivore and as a kairomone stimulating the feeding of other herbivores, depending upon the vagaries of mutations and coevolution. Specific kairomones may be synthesized by a wide variety of plant genera or families, and thus such kairomones may regulate host selection by insect herbivores that are monophagous, stenophagous, oligophagous, or polyphagous.

As our understanding of the intricacies of chemical ecology increases, it has become apparent that plant secondary compounds used by most insect herbivores as kairomones to promote host selection can in turn be sequestered by the herbivore and used as allomones to deter predation. Insect predators and parasitoids (the third trophic level) use kairomones of herbivores as clues to facilitate carnivory and parasitism.

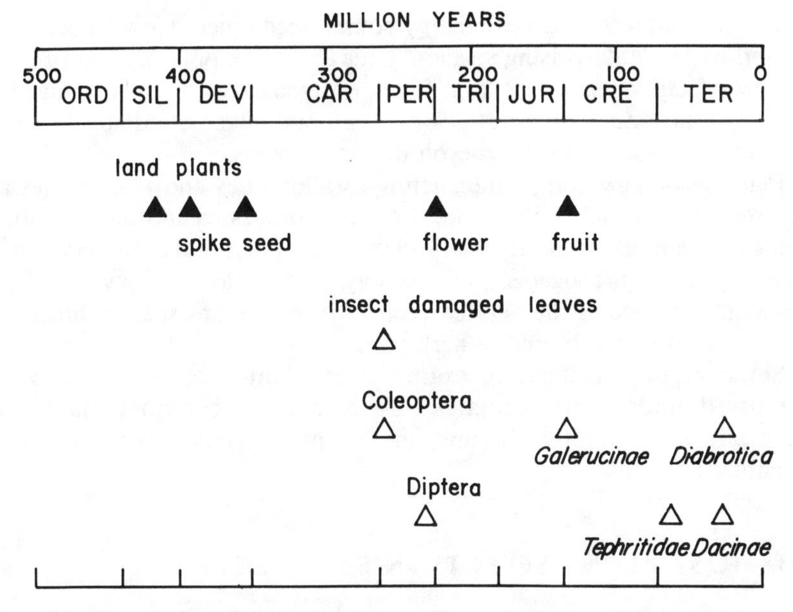

Figure 1.1. Coevolution of plants and insects. (Smart & Hughes 1973 and Riek 1970). Reprinted with permission from Metcalf (1985).

II. COEVOLUTION OF PLANTS AND INSECTS

Land plants first appeared at the beginning of the Devonian age, about 400 million years (myr) before the present (BP). Insects were not far behind, and the oldest insect fossil is that of a bristle-tail (Thysanura) *Gaspeya paloventognathae* found in lower Devonian rocks dated at about 390 myr BP (Labandeira et al. 1988). This species must have fed on early land plants. Thus for almost 400 million years, plants and insects have been engaged in a continuing struggle for ascendency in the terrestrial world. Some of the salient history of these interrelationships between the two dominant life forms, plants and insects, is shown in Figure 1.1 (Riek 1970, Smart & Hughes 1973). The plant-herbivore interface has been described by Ehrlich and Raven (1964) as the major zone of interaction responsible for generating the diversity of terrestrial life forms. These authors advanced the persuasive idea of coevolution to delineate the importance of reciprocal selective responses between closely linked organisms in generating terrestrial organic diversity.

As Price (1984) has characterized it, the essential preoccupation of all species is to acquire sufficient energy to facilitate reproduction, and this energy must be maintained against all competitors. If another organism higher in the trophic system can penetrate the defense system of a plant, it will have access to a new energy source (ecological niche) that is not shared by less enterprising species. Thus an attack on a plant by an herbivore selects for an adaptation for new defenses by the plant, and this cycle of attack and counterattack between organisms of adjacent trophic levels is the essence of the coevolutionary process.

Plant secondary compounds acting as allomones and kairomones are key regulating factors of the coevolutionary processes and these relationships between plants, chemicals, and insects are seminal in understanding insect speciation, zoogeography, ecology, and behavior. They are of primary importance in the applied ecology of host-plant susceptibility and resistance to attack by insect herbivores.

Some of the outstanding examples of plant-insect coevolution are discussed under glucosinolates (Chapter 1), hypericin (Chapter 1), cucurbitacins (Chapter 4), and the chemical ecology of pollination (Chapter 6).

III. HOST PLANT SELECTION BY INSECTS

Host plant selection by phytophagous insects is a complex process involving the sequential effects of plant produced semiochemicals on insect behavior through host finding, feeding, oviposition, and growth and development. In this process, plant semiochemicals influence insect behavior (Kogan 1976, 1982) by acting as:

allomones
repellents–orient insects away from plants
excitants–initiate or accelerate movement
suppressants–inhibit biting or piercing
deterrents–deter feeding or oviposition
antibiotics–disrupt normal growth and development
antixenotics–disrupt normal host selection
kairomones
attractants–orient insects toward plants
arrestants–slow down or stop insect movement
excitants–elicit biting, piercing, or oviposition

In this discussion we limit ourselves to the effects of plant kairomones (Greek *kairos*, opportunistic) on insect behavior.

Kogan (1976) has identified several modes of insect host selection behavior:

a) polyphagous insects such as grasshoppers and locusts (Orthoptera: Locustidae) that orient to food plants through photo- and anemotaxis, with green volatiles playing a key role in plant recognition. Host plant acceptance results from universal feeding stimulants such as sugars, amino acids, lipids, and vitamins. Oviposition is non-selective.

b) polyphagous insects, such as aphids and whiteflies (Homoptera) or thrips (Thysanoptera), that orient to plants by green-yellow-orange spectral reflectance and are arrested and stimulated to feed and oviposit by sugars and amino-acids detected by probing.

c) oligophagous insects such as rootworm beetles (Coleoptera: Chrysomelidae) where adults and larvae exhibit widely different habitus and host preferences. Specific phenylpropanoid attractants and triterpenoid arrestants determine adult feeding preference, but oviposition is non-selective and eggs are laid in non-botanical substrates.

d) oligophagous insects such as the Colorado potato beetle (Coleoptera: Chrysomelidae) and the tomato and tobacco hornworms (Lepidoptera: Sphingidae)that are restricted to the plant family Solanaceae. Host finding is probably through anemotaxis stimulated by green volatiles, and contact chemoreception of nicotine and solanidine alkaloids, that are usually feeding deterrents, define host acceptance and oviposition.

e) oligophagous species such as *Pieris*, *Plutella*, and *Autographa* (Lepidoptera), *Phaedon* and *Phyllotreta* (Coleoptera), and *Brevicoryne* (Homoptera) that are restricted to Cruciferae as hosts by attraction, arrest, and excitation to oviposition by sinigrin and related glucosinolates and the volatile metabolite allyl isothiocyanate. These kairomones are feeding stimulants for the larvae.

Generalizations about insect host plant selection are in constant need of modification as rapidly accumulating knowledge about chemical ecology demonstrates that insect behavior is highly dependent upon specific volatile and nonvolatile attractants, arrestants, and excitants.

IV. VOLATILE PLANT KAIROMONES AS INSECT ATTRACTANTS

A considerable range of plant semiochemicals has been identified that act as volatile attractants for associated insect species. Attractivity of these compounds to insects results from diffusion through air and is relatively long range. The semiochemicals analyzed in Table 1.1 act largely as kairomones to benefit the insect perceiver, although floral volatiles involved in pollination act as synomones benefiting both the emitting plant through pollination and the perceiving insect by rewards of nectar and pollen; or through more ecological rewards of aggregation or lek formation that lead to mating (Williams 1983, Chapter 6).

The examples of attractant semiochemicals listed in Table 1.1 have been selected to conform to Fraenkel's (1959) requirements for unequivocal demonstration of insect responses to semiochemicals: (1) isolation and identification of chemical structure, (2) initiation of a kairomonal response when the chemical is applied to a neutral substrate, and (3) demonstration of a quantitative dose-response. These well documented examples relate to more than 70 plant odorants that produce selective behavioral responses in more than 300 insect species of 5 orders. The odorants range from molecular weight 99 to 222, and in boiling point from 20 to 340° C. These parameters therefore define the physical properties of attractant kairomones.

The attractant kairomones of Table 1.1 are classified into chemical types as terpenoids (27), phenylpropanoids (20), alcohols (3), aldehydes (3), esters (8), acid (1), and sulfur compounds (2). The plant attractant is a single compound in a few well documented examples, such as methyl eugenol, for some 58 species of *Dacus* (Diptera: Tephritidae) fruit flies, and raspberry ketone for a related 176 species (Chapter 5), phenylacetaldehyde for moths (Lepidoptera: Noctuidae), and benzyl acetate, cineole, and eugenol for Euglossini bees (Hymenoptera: Apidae) (Chapter 6). However, in many examples based on the spectrum of volatiles present in blossoms and fruits, there is enhanced or synergistic attraction by mixtures of odorants, e.g. the Japanese beetle, *Popillia japonica* (Coleoptera: Scarabaeidae) (Chapter 3); the *Diabrotica* beetles (Coleoptera: Chrysomelidae) (Chapter 4); the elm bark beetles (Coleoptera: Scolytidae), and the apple maggot fly, *Rhagoletis pomonella* (Diptera: Tephritidae) (Chapter 5). Considering the pressures of evolution, optimum response to mixtures of semiochemicals would seem to be more probable, rather than exclusive reliance upon a single odorant.

A. Spectrum of Volatiles Produced by Plants

Recent studies by gas chromatography-mass spectrometry reported in *Phytochemistry* and in the *Journal of Chemical Ecology* have demonstrated that plant sources produce a rich menu of volatile compounds. For example, 40 volatiles have been isolated from *Cucurbita* blossoms (Andersen 1988), 46 from *Castania creata* chestnut flowers (Yamaguchi & Shibamoto 1980), 49 from the flowers of *Peony albiflora* (Kumar & Motts 1986), and 37 from the sunflower *Helianthus annuus* (Etievant et al. 1984). Fruits are equally prolific in the production of volatiles and more than 60 have been isolated from apples (Averill et al. 1988), 36 from *Psidium guajava* (MacLeod & de Tronconis 1982a), and 21 from *Mangifera indica* (MacLeod & de Tranconis 1982b). Other plant parts are

Table 1.1. Volatile Plant Kairomones Attractive to Insects.

Insect	Kairomone	Source	Mol. Wt.	Bp(°C)	Reference
		Lepidoptera			
Cisseps fulvicollis (yellow colared scape moth)	phenylacetaldehyde		120	195	Cantello & Jacobsen (1979a,b)
Heliothis zea (corn earworm and nine other Noctuidae)	phenylacetaldehyde	corn silk	120	195	ibid
Grapholitha molesta (oriental fruit moth)	terpineol acetate	peach	197		Brunson (1955)
Cydia pomonella (codling moth)	α-farnesene		204	ca.250	Wearing & Hutchins (1973)
Manduca quinquemaculata (tomato hornworm)	amyl benzoate		192		Morgan & Lyon (1928)
	amyl salicylate		208		
Manduca sexta (and 16 other Sphingidae)	amyl benzoate		192		ibid
	amyl salicylate		208		
Ostrinia nubilalis (European corn borer)	phenylacetaldehyde		120	195	Cantello & Jacobsen (1979a,b)
Papilio polyzenes asterius	*l*-carvone		150	230	Dethier (1941)
	anethole		148	234	
	anisaldehyde		136	246	
	anisic acid		152		
	estragole		148	216	

Table 1.1. (continued) Volatile Plant Kairomones Attractive to Insects.

Insect	Kairomone	Source	Mol. Wt.	Bp(°C)	Reference
		Coleoptera			
Acalymma vittatum (striped cucumber beetle)	indole	*Cucurbita* blossom	117	253	Andersen & Metcalf (1986)
					Lewis et al. (1990)
Anthonomus grandis (boll weevil)	β-bisabolol	*Gossypium* bud	222	ca.290	Gueldner et al. (1970)
	β-caryophyllene		204	ca.250	Hedin et al. (1975, 1976)
	β-caryophyllene oxide				
	limonene		136	175	Minyard et al. (1969)
Blastophagus piniperda (bark beetle)	α-pinene		136	155–156	Kangas et al. (1965)
	α-terpineol		154	217	
Carpophilus hemipterus (dried fruit beetle)	ethanol	fig	46	78–79	Smilanick et al. (1975)
	ethyl acetate		88	76–77	
	acetaldehyde		44	20–21	
Carpophilus mutilatus		fig			ibid
Cotinus nitida (green June beetle)	caproic acid	peach	116	202–203	
Dendroctonus pseudotsugae (Douglas fir beetle)	α-pinene	Douglas fir	136	155–156	Heikkenen & Hruitfiord (1965)
Diabrotica barberi (northern corn rootworm)	eugenol		164	254	Ladd et al. (1983)
	isoeugenol		164	266	Metcalf & Lampman (1989c)
	cinnamyl alcohol		134	250	
Diabrotica cristata	eugenol		164	254	Yaro et al. (1987)
	isoeugenol		164	266	Lampman & Metcalf (1988)
	cinnamyl alcohol		134	250	
Diabrotica undecimpunctata (southern corn rootworm)	cinnamaldehyde	*Cucurbita* blossom	132	246	Lampman et al. (1987)

Table 1.1. (continued) Volatile Plant Kairomones Attractive to Insects.

Insect	Kairomone	Source	Mol. Wt.	Bp(°C)	Reference
Diabrotica virgifera (western corn rootworm)	indole	*Cucurbita* blossom	117	253	Andersen & Metcalf (1986)
	estragole		148	216	
	β-ionone		192	ca.260	Lampman & Metcalf (1988)
Hylobius pales (pales weevil)	anethole		148	234	Thomas & Hertel (1969)
Hylurgopinus rufipes (native elm bark beetle)	α-phellandrene	elm	130	171	Millar et al. (1986)
	allo-aromadendrene		204	260	
	β-caryophyllene		204	ca.250	
	α-copaene		204	246	
	α-cubebene		204		
	β-elemene		204		
	α-gurjunene		204	ca.230	
	thujopsene		204	ca.250	
Leptinotarsa decemlineata (Colorado potato beetle)	*trans*-2-hexen-1-ol	potato	100	131	Visser & Ave (1978)
	cis-3-hexen-1-ol		100	131	
Listroderes costirostris (vegetable weevil)	coumarin		146	297	Matsumoto (1962) Matsumoto & Sugiyama (1960)
Popillia japonica (Japanese beetle)	anethole	grape, rose, peach	148	234	Wilde (1957)
	citronellol		156	224	Schwartz et al. (1966)
	eugenol		164	255	
	geraniol		154	229	
	phenethanol		122	219	
	caproic acid		116	202	
Phyllotreta cruciferae (cabbage flea beetle)	allyl isothiocyanate	Brassicae	99	148	Feeny et al. (1970)
Phyllotreta striolata (striped flea beetle)	allyl isothiocyanate		99	148	Feeny et al. (1970)

Table 1.1. (continued) Volatile Plant Kairomones Attractive to Insects.

Insect	Kairomone	Source	Mol. Wt.	Bp(°C)	Reference
Scolytus multistriatus (European elm bark beetle)	δ-cadinene	elm	204	ca.268	Meyer & Norris (1967)
	γ-cadinene		204	ca.260	Millar et al. (1986)
	±calamenene		202	255	
	α-cubebene		204		
	β-elemene		204	ca.240	
	α-muurolene		204		
Sitona cylindricollis (sweet clover weevil)	coumarin	sweet clover	146	297	Hans & Thorsteinson (1961)
		Diptera			
Ceratitis capitata (Mediterranean fruit fly)	terpineol acetate		196		Ripley & Hepburn (1935)
	α-copaene		204	246	Guiotto et al. (1972)
	α-ylangene		204		
Dacus cucurbitae (melon fly)	raspberry ketone		164	ca.340	Metcalf (1985, 1990) Drew (1989)
Dacus dorsalis (oriental fruit fly)	methyl eugenol		178	254	Howlett (1915) Metcalf (1990)
Delia antiqua (onion maggot fly)	dipropyl disulfide	onion	150	194	Matsumoto & Thorsteinson (1968) Dindonis & Miller (1980)
Delia brassicae (cabbage maggot fly)	allyl isothiocyanate	Brassicae	99	148	Finch (1978) Hawkes & Coaker (1979)
Psila rosae (carrot rust fly)	*trans*-asarone	carrot	208	296	Guerin et al. (1983)
	trans-2-hexenol		100	158	
	hexanal		100	131	
	heptanal		114	153	

Table 1.1. (continued) Volatile Plant Kairomones Attractive to Insects.

Insect	Kairomone	Source	Mol. Wt.	Bp(°C)	Reference
Rhagoletis pomonella	butyl 2-methylbutanoate	apple	160		Fein et al. (1982)
(apple maggot fly)	propyl hexanoate		158	187	
	butyl hexanoate		172	208	
	hexyl propanoate		158	190	
	hexyl butanoate		172	208	
		Hymenoptera			
Andrena spp.	γ-cadinene		204	ca.260	Priesner (1973)
Euglossa spp.	anisyl acetate		180		Williams & Whitten
	methyl salicylate		152	220	(1983)
	β-ionone		192		
	eugenol		164	254	
	1,8-cineole		154	176	
Eulaema bombiformis	benzyl acetate		150	213	Williams & Whitten (1983)
Eulaema cingulata	eugenol		164	254	Williams & Whitten (1983)
Euplussia spp.	1,8-cineole		154	176	Williams & Whitten (1983)
		Thysanoptera			
Frankliniella intonsa	anisaldehyde		136	248	Kirk (1985)
Thrips flavus					
T. major					
T. pillchi					
(flower thrips)					

well supplied and 68 volatiles have been identified from corn silk, *Zea mays* (Flath et al. 1978, Buttery et al. 1980), 33 from cabbage leaves, *Brassica oleracea* (MacLeod & Nussbaum 1977), and 16 from sugar beet leaves, *Beta vulgaris* (MacLeod et al 1981) (see Metcalf 1987). Such lists of plant volatiles can provide invaluable clues for the development of specific kairomones and mixtures for the attraction of phytophagous insects (Chapter 4, Section VI, Chapter 5, Section II-B).

The examples of plant odorants shown to be effective insect kairomones (Table 1.1) indicate that semiochemical odorants are generally restricted to compounds with molecular weights < 250 and boiling points $< 340°$ C. If we apply these limitations to the broad menu of secondary compounds present in plants, we are restricted to alcohols, aldehydes, and ketones of C_{16} or less (e.g. cetyl alcohol $C_{16}H_{32}OH$, mol wt 242, bp 344), acids of C_{14} or less (e.g. myristic acid $C_{13}H_{25}COOH$, mol wt 228, bp 250), esters of C_{14} or less (e.g. dodecyl acetate mol wt 228, bp 280), phenylpropanoids $< C_{12}$ (e.g. asarone mol wt 208, bp 296), and terpenoids $< C_{15}$ (e.g. cadinenes mol wt 204, bp 260). A large variety of secondary plant compounds are encompassed by these parameters, and it appears, from the variety of Table 1.1, that almost any one of these compounds, singly or in innumerable combinations, may be involved in the chemical ecology of plant/insect relationships.

B. Production of Volatile Kairomones by Plants

Plant volatiles attractive to insects as kairomones and synomones are typically lipophilic substances. Many of them that are especially agreeable and distinctive odorants of plant flowers and fruits have been characterized chemically as "essential oils". The terpenoids, derived by the condensation of mevalonic acid, are the most numerous and structurally varied of the essential oils. Those categorized as volatile odorants are classified by the number of 5-carbon isoprenoid units as: hemiterpenes (1 unit), monoterpenes (2 units), and sesquiterpenes (3 units) (Rodriguez et al. 1984). Many of these terpenoids are oxygenated in plants to a variety of alcohols, aldehydes, ketones, and esters. The chemical structures of terpenoids specifically identified as volatile insect kairomones (Table 1.1) are shown in Figure 1.2.

Phenylpropanoids comprise another important group of essential oils from plants. These have the generalized structure $C_6 \cdot C_3$ and are derived from the cyclization of erythrose-4-phosphate to shikimic acid with subsequent aromatization to phenylalanine (Geissman & Crout 1969, Friedrich 1976). The green volatiles are formed by the oxidative degradation of leaf lipids (Visser & Ave 1978).

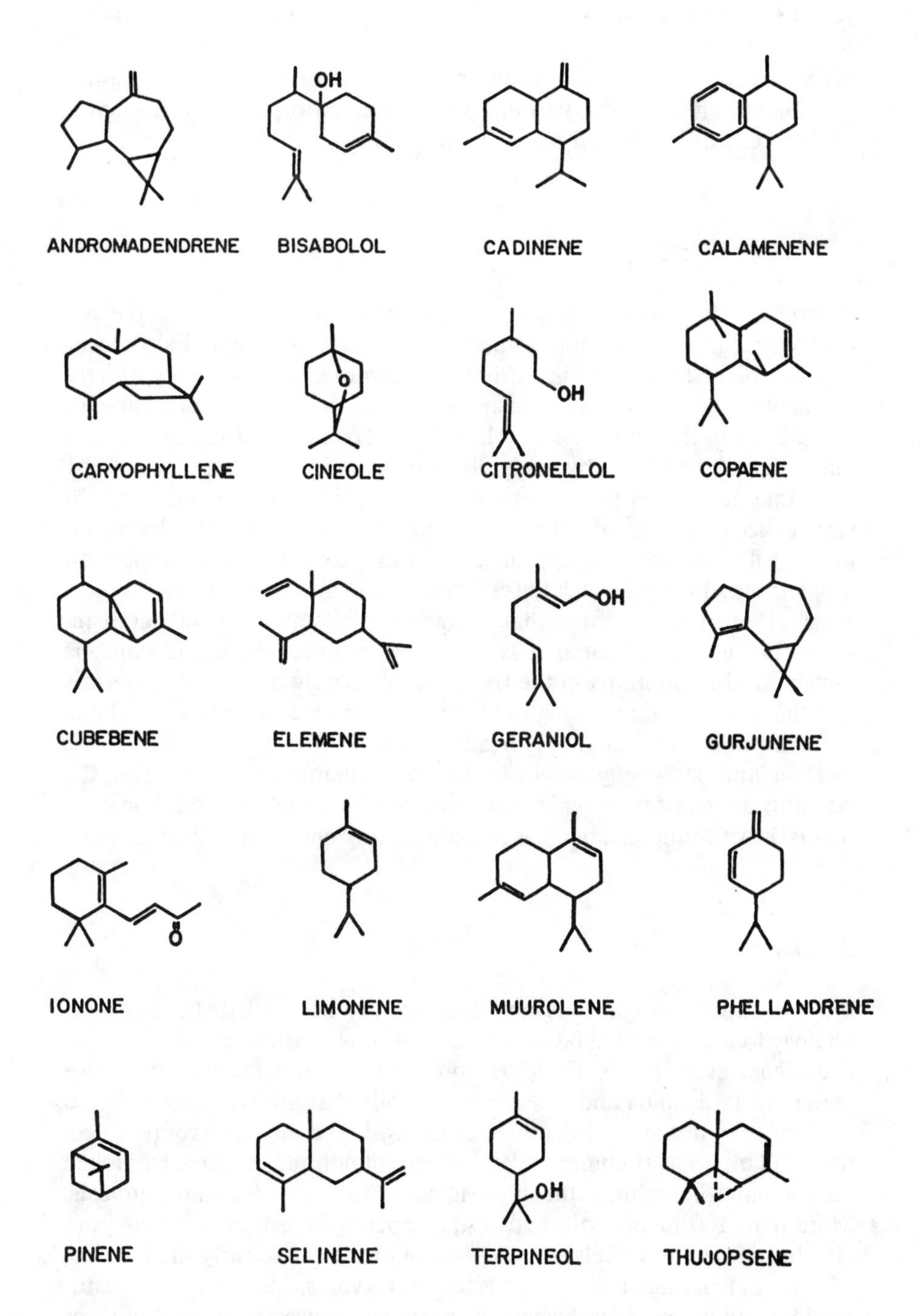

Figure 1.2. Plant terpenoids identified as insect kairomones (Reprinted with permission from Metcalf 1987.

Plants produce semiochemicals in four types of organs: osmophores, glandular trichomes of leaves and stems, ducted oil cavities or glands, and oil cells of leaves and fruits (Fahn 1979).

1. Osmophores

Osmophores are specific "scent" generating organs common to the Aristolochiaceae, Araceae, Burmaniaceae, and Orchidaceae and also occur in other plant families (Esau 1965). Osmophores have a highly diverse morphology of flaps, cilia, or brushes and reach their greatest evolutionary complexity in the Orchidaceae where the surface area of the osmophore may be greatly enlarged by finely divided mounds of rugose tissues or by numerous trichomes (Williams 1983). In orchids there is not any substantial accumulation of volatiles in the osmophores prior to release, but in other flowers odorants appear to form in the cytoplasm as minute oil droplets that diffuse through the epidermal layer and cuticle to the outside. Vogel (1966) has suggested that flowering plants accumulate starch reserves in the subepidermal cells of the osmophore where this starch is metabolized to odorants in the tissues immediately below the epidermis and these are released through the cuticle. The amounts of essential oils occurring in flower petals are typically small, 0.05% in *Jasminium*, 0.075% in *Rosa*, and 0.084% by weight in *Acacia* (Williams 1981). However, the odorants are constantly replaced as they volatilize, and the total production is 5 to 12 times as much as is found at any one time in fresh flowers.

2. Glandular Trichomes

These are highly specialized glandular secretory cells with distinctive morphology found in the Labiatiae, Solanaceae, Compositae, and Geranaciae (Rodriguez et al. 1984). These trichomes secrete and accumulate a large variety of terpenoids and other essential oils that are typically involved as allomones in the antimicrobial, antifungal, and antiherbivoral protection of plants. The trichomes take the form of pointed, hooked, and lobed hollow hairs containing the highest concentration of semiochemicals. More than 100 mono-, di-, and sesquiterpenes have been isolated from glandular trichomes (Kelsey et al. 1984) and they are clearly the first line of plant defense against pathogens and herbivores. Many of these plant-produced allomones have become kairomones in insect/plant relationships (Table 1.1).

3. Biosynthesis of Volatiles by Plants

Glucosinolates are among the most carefully studied plant kairomones, and are sulfur derivatives (Gl-S-C[R] = $NOSO_3$-) analogous to the cyanogenic glucosides (Kjaer 1960). Amino acids are the precursors of the more than 60 glucosinolates present in the Cruciferae, Capparidaceae, Resedaceae, Tovariaceae, Moringaceae, Limnanthaceae, Tropaeloaceae, Coricaceae, Euphorbiaceae, Gyrostemonaceae, and Savadoraceae (Ettlinger & Kjaer 1968). The presence of glucosinolates characterizes the distinctive odor and taste of the Cruciferae (Brassicaceae). Most of these plants also contain thioglucosidases that are rapidly released after damage to plant tissues, and that subsequently hydrolyze the glucosinolates to produce volatile isothiocyanates that are responsible for the pungent odors that characterize the Cruciferae. The glucosinolate sinigrin is present in the leaves of mustard, *Brassica nigra* from 0.18 to 0.66% fresh weight and is readily converted to the volatile allylisothiocyanate that has been shown to act as a kairomone for host selection by the flea beetles *Phyllotreta cruciferae* and *P. striolata*, and by the mustard beetle *Phaedon cochliariae* (Coleoptera: Chrysomelidae) (Feeny et al. 1970); by the vegetable weevil *Listroderes costirostris obliquus* (Coleoptera: Cucurlionidae); by the cabbage maggot fly *Delia* (= *Hylemya*) *brassicae* (Diptera: Anthyomyiidae) (Finch & Skinner 1982); by the cabbage worms *Pieris brassicae* and *P. rapae* (Lepidoptera: Pieridae); by the diamond-back *Plutella xylostella* (Lepidoptera: Plutellidae); by the cabbage looper *Trichoplusi ni* (Lepidoptera: Noctuidae); and by the cabbage aphid *Brevicoryne brassicae* (Homoptera: Aphidae) (Gornitz 1956). The glucosinolates represent the primary defensive allomones of a large group of related plant families of the order Capparales, and these glucosinolates have been exploited by a diversified group of herbivores as kairomones for host selection (Feeny 1976, Feeny et al. 1970). Detailed aspects of this kairomonal attraction of *D. brassicae* have been presented by Finch (1980).

4. Volatile Kairomones from Plant Senescence and Decay

Moribund or decaying plant tissues produce a spectrum of odorants that are significantly different from those released by healthy plants. The bark beetles (Coleoptera: Scolytidae) prefer to attack and colonize dying trees. Moribund elm trees, *Ulmus americana*, especially those attacked by the pathogenic fungus of Dutch elm disease *Ceratocystis ulmi*, have a sweet apple-like odor that is distinctively different from that of healthy trees (Millar et al. 1986). This odor is attributable to an increased concentration in the wood of the triterpenoid α-cubebene which is a kairomone released by moribund elms tissues infested with *C. ulmi* (Gore et al. 1977). The native American elm bark beetle, *Hylurgopinus rufipes*, a vector of Dutch elm disease, preferentially attacks moribund elms and is attracted to a mixture of at least seven sesquiterpenes (Table 1.1) released by decomposing elm wood (Millar et al. 1986). The oxidative decay products of lignin, vanillin and syringealdehyde are also attractive to the European elm bark beetle, *Scolytus multistriatus*, another vector of *C. ulmi* (Meyer & Norris 1967). This species is also attracted to a mixture of at least six terpenoids (Table 1.1) produced by decaying elm (Miller et al. 1986).

C. Release of Plant-Produced Kairomones

Key odorants are released from certain plant tissues by highly specialized glands such as osmophores and glandular trichomes. Air temperature is an important factor in the rates of production and release (e.g. orchid flowers produce little or no odorants on cool or overcast days or at night but release abundant odorants on warm, sunny days). Scent production in orchids is nicely correlated with Euglossini bee behavior (Chapter 6) as these insects are inactive during inclement weather and at night (Williams 1981). Plant species such as the Araceae, are said to produce heat that aids in the volatilization of odorants such as indole and skatole (Meeuse 1978).

Many plant produced kairomones stored in oil glands or at other sites are released by leakage. Volatile chemicals are leaked from plants in five major ways: (1) by diffusion through aerial and subterranean surfaces, (2) by leaching through the action of dew and rainwater, (3) by exudation, (4) by plant damage, and (5) by decay of plant material (Rice 1974).

V. NON-VOLATILE PLANT KAIROMONES AS INSECT ARRESTANTS, FEEDING STIMULANTS, AND OVIPOSITION STIMULANTS

A number of plant secondary compounds have been identified as non-volatile kairomones of relatively high molecular weight. These are active in modifying insect behavior as arrestants, feeding stimulants (phagostimulants), and oviposition stimulants. Examples where there is relatively complete information about isolation and chemical identification, initiation of kairomonal response when applied to a neutral surface, and demonstration of dose-response relationship, i.e. Fraenkel's (1959) postulates, are presented in Table 1.2. The insect responses to these kairomones present in plant tissues and on surfaces are produced by direct contact

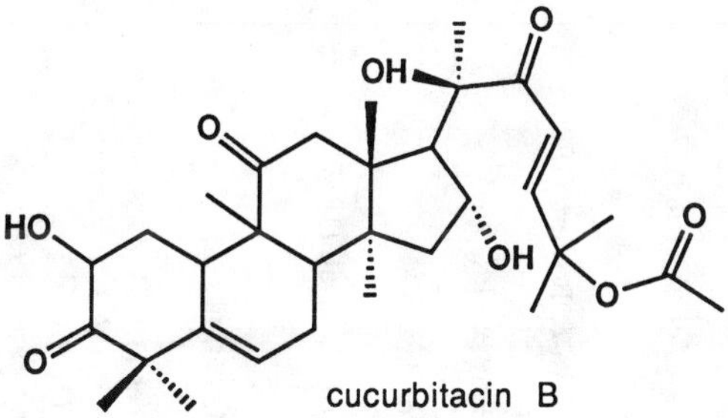

cucurbitacin B

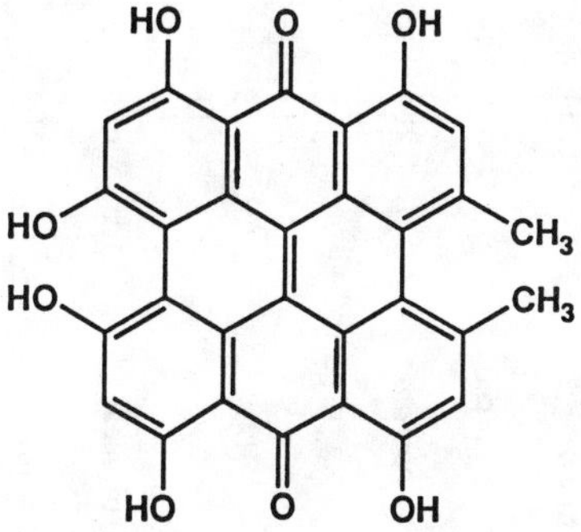

hypericin

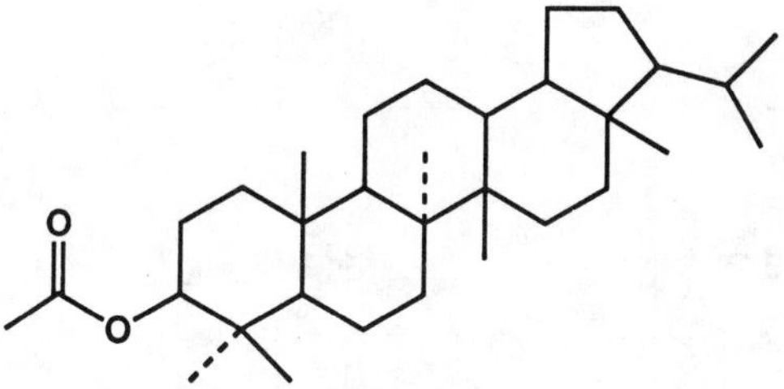

boehmeryl acetate

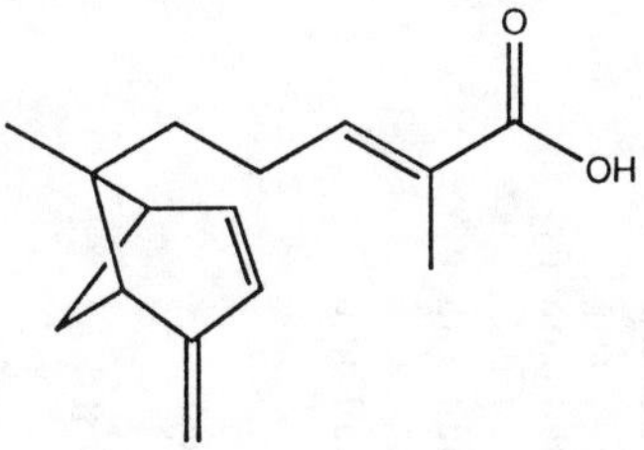

bergamoten-12-oic acid

Table 1.2. Non-Volatile Plant Chemicals as Insect Kairomones.

Insect	Kairomone	Source	Mol. Wt.	Reference
		Lepidoptera		
Atrophaneura alcinous (swallowtail butterfly)	aristolochic acid	*Aristolochia*	341	Nishida & Fukami (1989)
Ceratomia catalpae (catalpa sphinx larva)	catalposide	*Catalpa* spp.	482	Nyar & Fraenkel (1962)
Heliothis virescens (tobacco budworm moth)	α- and β-4,8,13-duvatrien-1-ols and 1,3-diols	*Nicotiana tabaccum*	306	Jackson et al. (1986)
Heliothis zea (corn earworm moth)	(*E*)-santalen-12-oic acid and (*E*)-*endo*-bergamaten-oic acid	*Lycopersicon hirsutum*	216 234	Coates et al. (1988)
Junonia coenia (buckeye butterfly)	catalpol	*Plantago lanceolata*	308	Pereya & Bowers (1988)
Papilio polyxenes (swallowtail butterfly)	luteolin	*Daucus carrota*	286	Feeny et al. (1988)
Papilio protenor (swallowtail butterfly)	chlorogenic acid, haringen, hesperidin	*Citrus*	342 611	Honda (1990)
Papilio xuthus (swallowtail butterfly)	nairutin, hesperidin, rutin, vicenin	*Citrus*	611	Nishida et al. (1987)
		Coleoptera		
Aulacophora (24 species)	cucurbitacin B,E	Cucurbitaceae	526, 524	Metcalf (1986)
Chrysolina (10 species)	hypericin	*Hypericum*	504	Rees (1969) Jolivet & Petitpierre (1976)
Cylas formicarius (sweet potato weevil)	boehmeryl acetate	*Ipomoea batatas*	468	Son et al. (1990)
Diabrotica (41 species)	cucurbitacin B,E	Cucurbitaceae	526, 524	Chambliss & Jones (1966) Metcalf (1986)
Epilachna varivestis (Mexican bean beetle)	phaseolutin lotaustrin	*Phaseolus* spp.		Klingenburg & Bucher (1960)
Leptinotarsa decemlineata (Colorado potato beetle)	chlorogenic acid	*Solanum tuberosum*	342	Hsio & Fraenkel (1968)

with chemoreceptors, usually located on the insect tarsi but sometimes present on maxillary palpi or ovipositor (Schoonhoven 1985, Stadler 1984).

Such non-volatile kairomones are definitive in regulating insect behavior and can be important in the applied ecology of insect pest control. There are possibilities in selective plant breeding to remove the production of oviposition stimulants from plant genomes and thus to improve host plant resistance by antixenosis. The arrestant and phagostimulant cucurbitacins have been shown to play important roles in applied insect control. Cucurbitacins regulate the feeding behavior and host selection by a large group of Luperini beetles that attack Cucurbitaceae. They are discussed in Chapter 4.

A. Hypericin and Biological Control of Klamath Weed

Hypericin, or 4,5,7,4′,5′,7′-hexahydroxy-2,2′dimethylnaphthodianthrone, is a crimson pigment with red fluorescence that is secreted by glands in the stems, leaves, and flowers of at least 140 species of *Hypericum* (Hyperiaceae) of the subgenera *Euhypericum* and *Campylosporus*, originating in Europe, North Africa, Asia Minor, and Asia (Mathis & Ourisson 1963). Vertebrate herbivores feeding on these plants suffer from intense photosensitization and skin irritation that sometimes lead to blindness and/or starvation. Thus hypericin is an allomone protecting these singular plants from attack by herbivores. Species of *Hypericum* are avoided by almost all herbivores. However, at least 12 species of *Chrysolina* beetles (Coleoptera: Chrysomelidae) are monophagous feeders on the weed species *H. perforatum*, *H. hirsutum*, *H. maculatum*, and *H. tomentosum* (Jolivet & Petitpierre 1976). The dominant factor in host preference by *Chrysolina* beetles is the presence of hypericin in the host plants where it occurs at concentrations ranging from 30–100 μg per g in stems and leaves, and up to 500μg per g in flowerheads (Rees 1969). The *Chrysolina* beetles have specific *sensilla chaetaca* receptors located on the pretarsi that respond to minute amounts of hypericin on leaf surfaces of *Hypericum* by inducing arrest and phagostimulation. Rees (1969) showed by electrotarsograms that hypericin produced depolarization of these receptors with a threshold response of 1×10^{-5} M and that there was a linear relationship between log dose and mean frequency of response up to 3×10^{-3} M. Homologous tarsal sensilla in *C. menthastri* and *C. polita*, two species that do not feed on *Hypericum*, were non-responsive to hypericin stimulation. The beetles *C. brunsvicensis* and *C. quadrigemina* feed avidly on *H. hirsutum*, *H. perforatum*, *H. teratperam*, *H. pulchrum* and *H. dubium* that contain hypericin, but not on *H. androsaemum* and *H. calycinum* that do not produce hypericin.

H. perforatum (St. John's wort or Klamath weed) was introduced into the Klamath River area of California in 1900 and by 1944 had infested an estimated 5 million acres of rangeland. Its poisonous properties greatly decreased the productivity and value of the rangeland and *Chrysolina quadrigemina* and *C. hyperici* beetles were introduced from Europe as biological control agents. By 1959, the biological control of Klamath weed in California was described as a complete success (Huffaker & Kennett 1959).

VI. DETECTION OF PLANT KAIROMONES BY INSECTS

Chemical ecology is the key factor involved in regulating the behavioral and ecological association between insects and plants. The many examples (Tables 1.1 and 1.2) of insects specifically attracted to semiochemicals produced by plants demonstrate the importance of olfaction and gustation in the processes of host selection. Comprehensive reviews of various aspects of this subject include: Dethier (1970), Schoonhoven (1972), Hedin et al. (1974), Kogan (1976), Finch (1980), Metcalf (1986, 1987, 1990), Metcalf & Lampman (1989a) and Visser (1986).

A. Insect Chemoreception

The insect antenna is usually the focal point of the receptive phase of insect/plant kairomone communication, although chemoreceptors for the non-volatile cucurbitacins are located on the maxillary palpi of the Diabroticite beetles (Chapter 4), and those for hypericin from *Hypericum* spp. are found on the foretarsi of *Chrysolina* spp. beetles (Rees 1969). Receptors for oviposition stimulants may be found on the female ovipositor.

Insect antennae are paired organs of the head innervated by the deutocerebrum of the brain. The jointed or filamentous antenna consists of three portions; the scape is the proximal portion attached to the head, the pedicel is short, and the flagellum or distal portion is usually long and filamentous but is sometimes reduced to a single segment (Schneider 1968). The olfactory functions of the antenna are revealed by its articulation and musculature permitting movement in all directions, by the nerve channels leading directly to association centers in the midbrain, and by the great variety of hairs, sensilla, and pore plates that typically cover the flagellum. The olfactory function of the antenna is most obvious

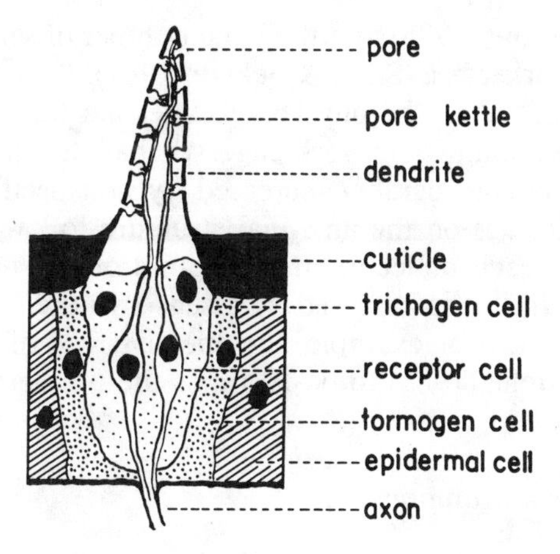

Figure 1.3. Diagram of typical insect olfactory receptor. (Reprinted with permission from Metcalf, 1987.

in the Lepidoptera, especially in the male giant silkworm moths (Saturniidae) whose plumose structure with many branches lined with filaments has evolved to intercept female sex pheromones at great dilutions in air, and thus to facilitate mating in species with very low population densities. Selective extirpation of the antennal segments in insects responsive to specific odorants decreases or abolishes the chemosensory response, and unilateral antennectomy profoundly disturbs the direction-finding mechanism involved in annemotactic behavior.

1. Structure of Olfactory Sensilla

Sensilla responsive to chemical stimuli occur in a variety of morphological forms, i.e. hair-like *sensilla trichodea*, cone-like *sensilla basiconica*, sensory pits *sensilla coeloconica*, and as pore-plates *sensilla placodea*. The olfactory function cannot be deduced from morphology alone, but a common feature is the presence of pores of 10–50 nm diameter, passing through the cuticle of the sensillum. These pores may number 3000–15,000 per sensillum and open into a labyrinth of tubes called "pore kettles" (Figure 1.3). The tubules connect with branching dendrites leading into the axon innervating the sensillum (Kramer 1978).

The sensitivity of the insect olfactory organ is determined by its efficacy in filtering out the odorant molecules from the air passing over the antenna. Mechanisms that have evolved to increase the antennal efficiency include the densely plumose antenna of the giant silkworm male moths

(Saturniidae) (Kramer 1978) and the large number of sensory pits on the antenna of the worker bee (Slifer & Sekhon 1961). Berg & Purcell (1977) have considered in detail the physics of molecular flow and diffusion in insect chemoreception. Their work suggests that the efficiency of insect chemoreception is considerably increased by nonspecific adsorption of the odorant molecules on the antennal structure followed by molecular diffusion, so that each molecule may rebound on the antenna as many as 250 times or until it is captured by a sensory pore. Thus, in the silkworm male *Bombyx*, for example, the effective size of the sensillum is estimated by Futrelle (1985) to be 10 times the geometric size.

2. Receptors for Kairomones

Semiochemicals convey stimuli and transmit behavioral messages only when they are received by and excite a suitable chemoreceptor. Therefore, the receptor concept is as fundamental to chemical ecology as it is to pharmacology. Receptors are defined by Williams (1986) as "distinct cellular constituents that are capable of recognizing subtle differences in the chemical composition of ligands". Receptors are membrane recognition sites and the receptor concept involves certain universal assumptions (Williams 1986):

a) ligand-receptor interaction is reversible
b) ligand-receptor association is bimolecular and dissociation is monomolecular
c) receptors of any specific type are equivalent and independent
d) receptor occupancy by a ligand does not alter free receptor opportunity or free ligand concentration
e) biological response results from an equilibrium between ligand and receptor
f) the receptor is a transducer that does not modify the ligand but is itself shifted from an inactive to an active state

From present day knowledge of insect receptors (Mustaparta 1990) it appears that the active-site of the receptor is a membrane-bound macromolecule that is complementary in size, shape, and stereochemical configuration to the stimulating chemical and to the position, number, and nature of its functional groups. Receptor specificity is produced by the complementarity of fit between the kairomone and the macromolecule, and results in a conformational change in the macromolecule that activates the receptor by opening ion channels that induce changes in conductance across cell membranes. This results in receptor depolarization and the receptor potential produced is transmitted to the brain. The more specialized the interaction between semiochemical and macromolecule, the lower the noise level of irrelevant stimulation and the higher the receptor sensitivity. It appears that the kairomone interacts with the external face of the receptor membrane, and that transduction and coding take place within the membrane. The allosteric-membrane hypothesis assumes: (1) a highly specific binding, (2) generation of the receptor potential by ligand-initiated conformational change, through an ion-gating mechanism of Na^+, K^+, or Ca^{++} ATP'ase, and (3) modulation of the receptor signal by a second messenger system of cAMP or cGMP (Dodd & Persaud 1981).

The binding energy between semiochemical and receptor macromolecule is extremely small, consistent with the millisecond decay times that characterize cell responses to chemicals, and with the need for reversibility of the semiochemical receptor complex. Types of chemical bonds typically involved in the complexing of ligands with macromolecular proteins are: (1) Van der Waals' forces, (2) hydrophobic interactions, (3) hydrogen bonds, (4) dipole-dipole interactions, (5) charge transfer complexes, (6) ion-dipole bonds, and (7) ionic bonds (Dodd 1976, Farmer 1980). Most semiochemicals have three appropriately located receptor binding groups and such three-site binding can explain the stereospecific interactions of all biologically active kairomones with their macromolecular receptors.

B. Electroantennographic Responses to Kairomones

Schneider (1955, 1957a, 1957b) devised the electroantennogram (EAG) to monitor the depolarization of antennal receptor cells of male silkworm moths, *Bombyx mori* exposed to varying quantities of the female sex pheromone bombykol or (*E,Z*)-10,12-hexadecadienol. In the Lepidoptera, EAG responses to sex pheromones are extraordinarily sensitive and that of the male silkworm moth was found to be essentially linear with the log of the bombykol concentration from 10^{-3} to 100 $\mu g\ cm^{-3}$ of air. For this species, the EAG threshold is about 1×10^7 molecules of bombykol cm^{-3} of air, as compared to a behavioral threshold of about 2×10^2 molecules cm^{-3} (Schneider et al. 1967).

The EAG response measures the summation of the receptor potentials from many olfactory cells of the antenna (Boeckh et al. 1965), but does not provide information about the number of olfactory receptors activated by a given semiochemical. Thus EAG responses are most useful in conjunction with behavioral observations on species responses to given olfactory stimulants, and can be used to compare the responses of various species to a common odorant.

Most host plant leaves, fruits, and blossoms contain from 30 to 80 or more volatile components comprising a wide range of chemical types and structures. The EAG technology is most useful in determining the odorants active in receptor depolarization and in measuring synergistic effects of mixtures. Fein et al. (1982) isolated and identified various apple fruit volatiles by GC-MS and demonstrated that propyl hexanoate was the most effective in EAG depolarization. They were able to prepare a mixture of apple volatile components (Table 1.1) that was fully as attractive as the natural apple extract (see Chapter 4). Andersen & Metcalf (1986) evaluated the EAG responses from 13 volatile fractions of *Cucurbita maxima* blossoms to the spotted cucumber beetle adult, *Diabrotica undecimpunctata howardi* and found that indole produced the maximum EAG response. Field bioassay showed that *D. u. howardi* was not attracted to indole at any concentration, but the related *D. v. virgifera* and *Acalymma vittatum* beetles were attracted to sticky traps baited with indole in a dosage-dependent way. Jang et al. (1989) studied the EAG responses of the Mediterranean fruit fly *Ceratitis capitata* to the stereoisomers of the parakairomone trimedlure (see Chapter 4). The male EAG responses were much greater than those of the females. However, the order of male fruit fly EAG response was isomer $C > B_1 > A > B_2$ as compared to the field attractancy of $C > A > B_1 > B_2$. These experiences illustrate one of the pitfalls of EAG experimentation, i.e. that a positive EAG response does not necessarily reflect the behavioral response of the intact insect.

The EAG responses of 7 species of *Yponomeuta* (Lepidoptera: Yponomeutidae) to leaf extracts from 6 host plants were compared with EAG responses to 57 plant varieties in a monumental study by Van der Pers (1981). The species studied and their principal host plants were:

Insect	Principal Host Plant	Host Family
Yponomeuta cagnagellus	*Euonymus europaeus*	Celastraceae
Y. evonymellus	*Prunus padus*	Rosaceae
Y. malinellus	*Malus* spp.	Rosaceae
Y. mahalebellus	*Prunus mahaleb*	Rosaceae
Y. padellus	*Crataegus Prunus* spp.	Rosaceae
Y. plumbellus	*Euonymus europaeus*	Celastraceae
Y. vigintipunctatus	*Sedum telephyum*	Crassulaceae

All of the *Yponomeuta* species showed a generalized high EAG response to the "green volatiles": hexan-1-ol, hexanal, *trans*-2-hexenal, *trans*-2-hexen-1-ol, *cis*-2-hexen-1-ol, *cis*-3-hexen-1-ol, and *cis*-3-hexenyl acetate.

It was concluded that this general group of volatiles plays a role in host finding by many phytophagous insects. However, the individual species of leaf roller moths responded to specific odorants, e.g. *Y. evonymellus* and *Y. plumbellus* to benzyl acetate, *Y. cagnagellus* and *Y. plumbellus* to nerol, and *Y. vigintipunctatus* to β-ionone. Cluster analysis of the EAG results for 57 plant volatiles to all 7 *Ypononeuta* species demonstrated that a relatively small dissimilarity was present between (1) *trans*-2-hexen-1-ol and *trans*-3-hexen-1-ol; (2) hexanoic, heptanoic, and octanoic acids; (3) heptanal and *trans*-2-hexanal; and (4) isomers of linalool. Two small groups of odorants, (5) benzaldyhyde, salicylaldehyde, *d*-limonene, α-phellandrene, and α-terpinene, and (6) 2-pentanone, 2-methyl-1-butanol, 3-methyl-1-butanol, 1-penten-3-ol, and 5-hexen-1-ol were also distinguished by low dissimilarity. There were no essential differences in the responses of male and female moths. Comparisons of the EAG responses of the oligophagous *Yponomeuta* sp. with that of the polyphagous leaf roller *Adoxaphyes orana* showed little dissimilarity, as might be expected from the many food sources they have in common. These data demonstrate that olfactory receptor sensitivities are tuned to food plant odors, and that antennal sensilla responsive to the complex of green leaf volatiles play a part in host plant finding for phytophagous insects. Sensilla responsive to specific volatiles produced by preferred host plants determine the individual peculiarities of host plant selection.

1. EAG Responses to Green Leaf Volatiles

Although from 30 to 50 volatiles are commonly identified in green plant tissues, a small group of C_6 alcohols, aldehydes, and esters have been characterized as "green leaf volatiles" that provide a distinctive odor background producing well-defined EAG responses in plant feeding Coleoptera, Lepidoptera, Diptera, and Homoptera (Visser 1983). It appears that the overall chemotactic response to the spectrum of these volatiles is a generalized mechanism of host plant recognition by phytophagous insects. The EAG responses to the "green leaf volatiles" have been most thoroughly studied in the responses of the Colorado potato beetle, *Leptinotarsa decemlineata* to the volatiles from the potato *Solanum tuberosum* (Visser 1979). Single cell recordings from the antennal *sensilla basiconica* have differentiated 5 types of receptor responses to: *trans*-2-hexen-1-ol, *cis*-3-hexen-1-ol, hexan-1-ol, *trans*-2-hexenal, hexanal and *cis*-3-hexenyl acetate, and *trans*-3-hexen-1-ol and *cis*-2-hexen-1-ol (Ma & Visser 1978, Visser 1979). It was concluded that the potato beetle has a well developed sensory capacity for discrimination among the green leaf volatiles and that host selection in this stenophagous species relates to the total olfactory pattern experienced.

The olfactory response of the migratory locust, *Locusta migratoria*, has also been evaluated by recording from single *sensilla coeloconica* (sensory pits) (Kafka 1970, 1974). The predominant receptor response was activated by C_5-C_9 aliphatic alcohols, acids and aldehydes. However, distinct reaction groups or key stimulating substances could not be identified.

2. EAG Responses to Phenyl Propanoids

Phenyl propanoids are present as trace volatiles in a great variety of plants, and these compounds such as estragole, anethole, eugenol, isoeugenol, methyl eugenol, methyl isoeugenol, elemicin, and asarone are not only important flavor components of leaves and fruits but have also been selected evolutionarily as host recognition cues by a variety of insects (Table 1.1). The carrot fly *Psila roseae* exhibits a strong EAG response to *trans*-asarone and *trans*-methyl isoeugenol present in carrot leaves, *Daucus carrota* (Guerin et al. 1983). The response is very specific to the individual phenyl propanoids (Figure 1.4) and is maximal with the propenylbenzenes having the *trans*-configuration, and with 3,4,6-trimethoxy substitution. *Trans*-asarone, or 3,4,6-trimethoxy-l-propenylbenzene, produced an EAG response 3.7 times that of *cis*-asarone and 10.7 times greater than *trans*-methyl isoeugenol or 3,4-dimethoxy-l-propenyl benzene. Isoeugenol was inactive and methyl eugenol with the 2,3 − C = C side chain was less than 0.5 times as active as *trans*-methyl isoeugenol with the 1,2 − C = C side chain (Guerin et al. 1983). This EAG data emphasizes the specificity of the antennal receptor site for a limited range of phenyl propanoids and indicates that the overall sensitivity of the EAG response by the carrot fly to *trans*-asarone, with a threshold of about 10^{-5} μg, is comparable to that of the male silkworm moth to the sex pheromone bombykol (Schneider et al. 1967).

C. Specificity of Chemoreceptor Response

Most of the semiochemicals regulating insect/plant interactions convey their messages by impinging on specific receptor cells that are evolutionarily attuned to respond to highly specific molecular configurations. These olfactory or gustatory receptors are "specialist receptors" and there is no evidence that truly "generalist receptors" exist in insects. Clearly, the more specific this interaction between semiochemical and receptor, the lower the "noise level" from irrelevant receptor stimulation, and conse-

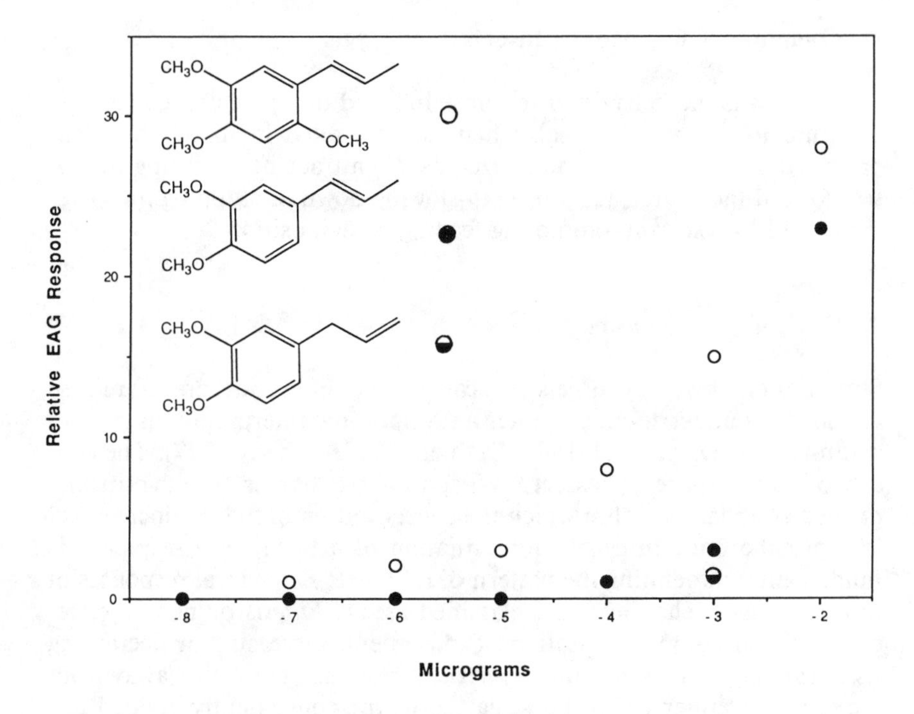

Figure 1.4. Electroantennographic (EAG) responses of the carrot fly, *Psila rosae*, to the plant kairomones asarone, methyl isoeugenol and methyl eugenol. (Data from Guerin and Visser, 1980).

quently the greater the receptor sensitivity. As a generalization, it appears that there is greater latitude in kairomone receptor specificity than there is in pheromone receptor specificity. Pheromone receptors often exhibit extremes of structural specificity as, for example, in the male silkworm moth *Bombyx mori* where the threshold response of the male to the pheromone (*E*,*Z*)-10, 12-dodecadienol is about 10^{-5} μg compared to 10^{-2} for the (*Z*,*E*)-isomer and 10^{-1} for the (*E*,*E*)-isomer (Schneider et al. 1967). Nevertheless, kairomone receptors exhibit impressive selectivity of responses. Thus the onion maggot fly, *Delia antiqua* responds to dipropyl disulfide but not to dipropyl sulfide (Dindonis & Miller 1980), and the threshold electroantennographic response of the carrot fly *Psila rosae* to asarone or 3,4,6-trimethoxy-1-propenylbenzene is about 10^{-7} compared to 10^{-4} for methyl isoeugenol or 3,4-dimethoxy-1-propenylbenzene (Guerin & Visser 1983). Receptor specificity of Dacinae fruit flies is discussed in detail in Chapter 5 and that of Diabroticite beetles in Chapter 4.

D. Chemotactic Behavior of Insects

Chemotaxis is the behavioral response induced in a perceiving insect by exposure to a semiochemical. Chemotaxes evoked in insects by plant produced kairomones are categorized as: (1) attractant–orienting the insect toward the source, (2) arrestant–slowing down or stopping its movement, and (3) excitant–promoting feeding or oviposition.

1. Guidance Mechanisms

Strategies employed by insects to locate sources of semiochemical release are not fully understood, and their anthropogenic interpretations remain controversial (David et al. 1982, Dethier 1970, Kennedy 1977). The concept of "active space", (Bossert & Wilson 1963) (Chapter 2), with critically defined boundaries within which the concentration of the semiochemical is above the odor threshold for initiation of a behavioral response, is fundamental in defining the pattern of response. Behavioral responses of insects to semiochemicals are classified as: (1) *kinesis* or nondirect response in initiation or cessation of movement, increasing or decreasing its linear rate of turning, and (2) *taxis*, a response toward or away from the source (Dethier 1970). These categories are conveniently divided further: kinesis may be either a change in the rate of linear motion, *orthokinesis*, or a change in rate or frequency of turning or *klinokinesis*. Similarly, taxis may be directed by *tropotaxis*, seeking equal stimuli from bilaterally symmetrical sensors, e.g. the paired antenna, or by *klinotaxis*, sampling with a single sensor at various times and different loci (Kennedy 1977). The *tropotactic* response seeks to keep the stimulations of both antennae constant by turning toward the more strongly stimulated side, and the feedback mechanism involved causes the insect to oscillate back and forth across the odor corridor. Chemotaxis is generally believed to be effective only when there are sharply defined odor gradients, especially near the source. The oriental fruit fly, *Dacus dorsalis*, responds to the plant kairomone methyl eugenol with a direct unwavering tropotactic advance, but unilateral antennectomy caused a spiraling *klinotactic* approach in contrast to the normal behavior (Metcalf et al. 1979).

2. Anemotaxis

Anemotaxis is orientation of an insect with respect to the direction of the wind current (Kennedy 1977). The presence of a semiochemical at concentrations equal to or above the behavioral threshold, impinging on the appropriate antennal receptor stimulates tactic movement upwind that follows a zigzag or "hunting" pattern. When the insect inadvertently veers from the active odor space, it performs a series of abrupt changes of course, often increasing flight direction to almost right angles to the wind direction. These hunting maneuvers are believed to be an efficient way to reestablish contact with the odor corridor. The final approach to the source of the semiochemical is facilitated by direct chemotaxis in response to the odor gradient present. This combination of maneuvers has been identified in the pollination activities of solitary bees with orchid flowers and for the short range orientation of the male bee in pseudocopulation with the pollinating apparatus of the orchid (Chapter 6).

Reversed anemotaxis occurs when insects inadvertently fly out of the active space and chemotaxis is abruptly terminated. In this situation, the typical response of a flying insect is to initiate zigzag movements over a wide range until the active space is reentered. Kennedy (1977) has suggested that the anemotactic response to semiochemical stimulation requires a highly integrated involvement of the central nervous system. The foraging honeybee utilizes a modified system of anemotactic behavior to maintain a direct track between hive and blossom by turning its body axis to adjust its course to the speed and direction of the wind. It is believed that this course is plotted by holding the angle of the ground track constant to the angle of the sun (Gould 1974, von Frisch 1967, Wenner 1974).

REFERENCES

Andersen, J.F. 1988. Composition of the floral odor of *Cucurbita maxima* Duchesne (Cucurbitaceae). J. Agr. Food Chem. 35: 60–62.

Andersen, J.F. and R.L. Metcalf. 1986. Identification of a volatile attractant for *Diabrotica* and *Acalymma* spp. from blossoms of *Cucurbita maxima* Duchesne. J. Chem. Ecol. 12: 687–699.

Andersen, J.F. and R.L. Metcalf. 1987. Factors influencing the distribution of *Diabrotica* spp. in the blossoms of cultivated *Cucurbita* spp. J. Chem. Ecol. 12: 681–699.

Averill, A.L., W.H. Reissig and W.L. Roelofs. 1988. Specificity of olfactory response in the tephritid fruit fly *Rhagoletis pomonella*. Entomol. Exp. Appl. 47: 211–222.

Berg, H.C. and E.M. Purcell. 1977. Physics of chemoreception. Biophys. Jour. 20: 193–219.

Boeckh, J., K.E. Kaissling and D. Schneider. 1965. Insect olfactory receptors. Cold Spring Harbor Symp. Quant. Biol. 30: 263–280.

Bossert, W.H. and E.O. Wilson. 1963. The analysis of olfactory communication among animals. J. Theoret. Biol. 5: 443–469.

Brunson, M.H. 1955. Effect on the Oriental fruit moth of parathion and EPN applied to control the plum curculio on peach. J. Econ. Entomol. 48: 390–392.

Buttery, R.G., L. Ling and R. Teraneshi. 1980. Volatiles of corn tassels: possible corn earworm attractants. Jour. Agr. Food Chem. 28: 771–774.

Cantello, W.W. and M. Jacobsen. 1979a. Cornsilk volatiles attract many species of moths. J. Environ. Sci. Health 14: 695–707.

Cantello, W.W. and M. Jacobsen. 1979b. Phenylacetaldehyde attracts moths to bladder flower and to blacklight traps. Environ. Entomol 8: 444–447.

Chambliss, O.L. and C.M. Jones. 1966. Cucurbitacins: specific insect attractants from Cucurbitaceae. Science 153: 1392–1393.

Coates, R.M., J.F. Denissen, J.A. Juvik and B.A. Babka. 1988. Identification of α-santolenic acid and *endo*-β-bergamotenic acids as moth oviposition stimulants from wild tomato leaves. J. Org. Chem. 53: 2186–2190.

Croteau, R. and M.A. Johnson. 1984. Biosynthesis of terpenoids in glandular trichomes, pp. 133–185 in E. Rodriguez, P.L. Healey and I. Mehta, eds. "Biology and Chemistry of Plant Trichomes". Plenum Press, N.Y.

David, C.T., J.S. Kennedy, A.K. Ludlow, J.N. Perry and C. Wall. 1982. A reappraisal of insect flight towards a distant point source of wind-borne odor. J. Chem. Ecol. 8: 1207–1215.

Dethier, V.G. 1941. Chemical factors determining the choice of food plants by *Papilio* larvae. Am. Nat. 75: 61–73.

Dethier, V.G. 1970. Chemical interactions between plants and insects, pp. 83–102 in E. Sondheimer and J.B. Simeone "Chemical Ecology", Academic Press, N.Y.

Dethier, V.G. 1980. Mechanisms of host plant recognition. Entomol. Exp. Appl. 31: 49–56.

Dethier, V.G., L.B. Browne and C.N. Smith. 1960. The designation of chemicals in terms of the responses they elicit from insects. J. Econ. Entomol. 53: 134–136.

Dindonis, L.L. and J.K. Miller. 1980. Host-finding responses of onion and seed corn flies to healthy and decomposing onions and several synthetic constituents of onion. Environ. Entomol. 9: 467–472.

Dodd, G.H. 1976. Ligand binding phenomena in chemoreception, pp. 55–63 in G. Benz ed. "Structure-Activity Relationships in Chemoreception." Information Retrieval Ltd., London.

Dodd, G. and K. Persaud. 1981. Biochemical mechanisms in vertebrate primary olfactory neurons. Chapt. 16 in R.H. Cagan & M.R. Kave, eds. "Biochemistry of Taste and Olfaction." Academic Press, N.Y.

Drew, R.A.I. 1989. The tropical fruit flies (Diptera: Tephritidae: Dacinae) of the Australasian and Oceanean Region. Mem. Queensland Mus. 26, South Brisbane, Australia.

Ehrlich, P.R. and P.H. Raven. 1964. Butterflies and plants, a study in coevolution. Evolution 18: 586–608.

Esau, K. 1965. "Plant Anatomy." 2nd. ed. John Wiley & Sons, N.Y.

Etievant, P.X., M. Azar, M.H. Pham-Delegue, and C.J. Masson. 1984. Isolation and identification of volatile constituents of sunflowers (*Helianthus annuus* L.). J. Agric. Food Chem. 32: 503–509.

Ettlinger, M.G. and A. Kjaer. 1968. Sulfur compounds in plants. Rec. Adv. Phytochem. 1: 59–144.

Fahn, A. 1979. Secretory Tissues in Plants, Academic Press, London.

Farmer, P.S. 1980. Bridging the gap between bioactive peptides and non-peptides: some perspectives in design, pp. 119–143 in E.J. Ariens, ed. "Drug Design." Vol. 10, Academic Press, N.Y.

Feeny, P. 1976. Plant apparency and chemical defense, in "Biochemical Interaction between Plants and Insects", Rec. Adv. Phytochem. 10: 1–40.

Feeny, P., K.L. Paauwe and N.J. Demong. 1970. Flea beetles and mustard oils: host plant specificity in *Phyllotreta cruciferae* and *P. striolata* adults (Coleoptera: Chrysomelidae). Ann. Entomol. Soc. Am. 63: 832–841.

Feeny, P., K. Sachdev, L. Rosenberry and M. Carter. 1988. Luteolin 7-*O*-(6'-*O*-malonyl)-β-

D-glucoside and *trans*-chlorogenic acid: oviposition stimulants for the black swallowtail butterfly. Phytochemistry 27: 3439–3448.

Fein, B.L., W.H. Reissig and W.L. Roelofs. 1982. Identification of apple volatiles attractive to the apple maggot, *Rhagoletis pomonella*. J. Chem. Ecol. 8: 1473–1487.

Finch, S. 1978. Volatile plant chemicals and their effect on host plant finding by the cabbage root fly (*Delia brassicae*). Entomol. Exp. Appl. 24: 350–357.

Finch, S. 1980. Chemical attraction of plant-feeding insects to plants, pp. 67–143 in T.H. Coaker, ed. "Applied Biology." Vol. 5, Academic Press, N.Y.

Finch, S. and G. Skinner. 1982. Trapping cabbage root flies in traps baited with plant extracts and with natural and synthetic isothiocyanates. Entomol. Exp. Appl. 31: 133–139.

Flath, R.A., R.R. Forrey, J.O. John and B.C. Chan. 1978. Volatile components of corn silk; possible corn earworm attractants. J. Agric. Food Chem. 26: 1290–1293,

Fraenkel, G. 1959. The raison d'être of secondary plant substances. Science 129: 1466–1470.

Fraenkel, G. 1969. Evaluation of our thoughts on secondary plant compounds. Entomol. Exp. Appl. 12: 473–486.

Friedrich, H. 1976. Phenylpropanoid constituents of essential oils. Lloydia 39: 1–7.

Futrelle, R.P. 1985. How molecules get to their detectors. The physics of diffusion of insect pheromones. Trends Neurosci. 8: 116–117.

Geissman, T.A. and D.H.G. Crout. 1969. "Organic Chemistry of Secondary Plant Metabolism", Freeman, Cooper & Co., San Francisco, CA.

Gore, W.E., G.T. Pearce, G.N. Lanier, J.B. Simeone, R.M. Silverstein, J.W. Peacock and R.A. Cuthbert. 1977. Aggregation attractant of the European elm bark beetle *Scolytus multistriatus*. Production of individual components and related aggregation behavior. J. Chem. Ecol. 3: 429–446.

Gornitz, K. 1956. Weitere Untersuchungen uber Insekten-Attraktiostoffe aus Cruciferen. Nachrichtenblatt. Dtsch. Pflanzenschutzdienst N.F. 10: 137.

Gould, J.L. 1974. Honey-bee communication. Nature (London) 252: 300.

Gueldner, R.C., A.C. Thompson, D.D. Hardee and P.A. Hedin. 1970. Constituents of the cotton bud. XIX. Attractancy to the boll weevil of the terpenoids and related plant constituents. Jour. Econ. Entomol. 63: 1819–1821.

Guerin, P.M., E. Stadler and H.R. Buser. 1983. Identification of host plant attractants for the carrot fly *Psila rosae*. J. Chem. Ecol. 9: 843–861.

Guerin, P.A. and J.H. Visser. 1983. Electroantennogram responses of the carrot fly to volatile plant components. Physiol. Entomol. 5: 111–119.

Guiotto, A., U. Fornasiero and F. Baccichetti. 1972. Investigation of attractants for males of *Ceratitis capitata*. Farmaco ed. Sci. 27: 663–669.

Hans, H. and A.J. Thorsteinson. 1961. The influence of physical factors and host plant odor on the induction and termination of dispersal flights in *Sitona cylindricollis* Fahr. Entomol. Exp. Appl. 4: 165–177.

Hawkes, C. and T.H. Coaker. 1979. Factors affecting the behavioral response of the adult cabbage root fly, *Delia brassicae*. Entomol. Exp. Appl. 25: 45–58.

Hawkes, C., S. Patton and T.H. Coaker. 1978. Mechanisms of host plant finding in adult cabbage root fly, *Delia brassicae*. Entomol. Exp. Appl. 24: 219–227.

Hedin, P.A., F.G. Maxwell and J.H. Jenkins. 1974. Insect plant attractants, feeding stimulants, repellents, deterrents, and other related factors affecting insect behavior, pp. 494–527 in F.G. Maxwell and F.A. Harris, eds. "Proceedings of Summer Institute of Biological Control of Plants, Insects and Diseases." Univ. Press, Jackson, Miss.

Hedin, P.A., A.C. Thompson and R.C. Gueldner. 1975. Survey of air space volatiles of the cotton plant. Phytochemistry 14: 2088–2090.

Hedin, P.A., A.C. Thompson and R.C. Gueldner. 1976. Cotton plant and insect constituents that control bollweevil behavior and development. Recent Adv. Phytochem. 10: 271–350.

Heikkenen, H.J. and B.F. Hruitfiord. 1965. *Dendroctonus pseudotsugae:* a hypothesis regarding its primary attractant. Science 150: 1457–1459.

Honda, K. 1990. Identification of host-plant chemicals stimulating oviposition by the swallow-tail butterfly *Papilio protenor.* J. Chem. Ecol. 16: 325–337.

Howlett, F.M. 1915. Chemical reactions of fruit flies. Bull. Entomol. Res. 6: 297–305.

Hsaio, T.H. and G. Fraenkel. 1968. Isolation of phagostimulative substances from the host plant of the Colorado potato beetle. Ann. Entomol. Soc. Amer. 61: 476–484.

Huffaker, C.B. and C.E. Kennett. 1959. A ten year study of vegetational changes associated with biological control of Klamath weed. J. Range Management 12: 69–82.

Jackson, D.M., R.F. Severson, A.W. Johnson and G.A. Herzog. 1986. Effects of cuticular divane diterpenes from green tobacco leaves on tobacco budworm (Lepidoptera: Noctuidae) oviposition. J. Chem. Ecol. 12: 1349–1359.

Jang, E.B., D.M. Light, J.C. Dickens, T.P. McGovern and J.T. Nagota. 1989. Electroantennogram responses of Mediterranean fruit fly *Ceratitis capitata* (Diptera: Tephritidae) to trimedlure and its isomers. J. Chem. Ecol. 15: 2219–2231.

Jolivet, P. and E. Petitpierre. 1976. Les plantes-hôtes connues des *Chrysolina* (Col. Chrysomelidae) essai sur les types de sélection trophique. Ann. Soc. Ent. Fr. (N.S.) 12: 123–149.

Kafka, W.A. 1970. Moleculare Wechselwirkung bei der Erregung einzeiner Reichgelten. J. Comp. Physiol. 70: 105–143.

Kafka, W.A. 1974. Physiocochemical aspects of odor reception in insects. Ann. N.Y. Acad. Sci. 237: 115–128.

Kangas, E., V. Pertunnen, H. Oksanen and M. Rinne. 1965. Orientation of *Blastophagus piniperda* L. to its breeding material. Entomol. Fennici 31: 61–73.

Kelsey, R.G., G.W. Reynolds and E. Rodriguez. 1984. The chemistry of biologically active constituents secreted and stored in plant glandular trichomes, pp. 187–241, in E. Rodriguez, P.L. Healey and J. Mekta, eds "Biology and Chemistry of Plant Trichomes." Plenum Press, N.Y.

Kennedy, J.S. 1977. Olfactory responses to distant plants and other odor sources, Chapter 5 in H.H. Shorey and J.H. McKelvey, eds. "Chemical Control of Insect Behavior." John Wiley & Sons,N.Y.

Kirk, W.D.J. 1985. Effect of some floral scents on host finding by thrips (Insecta: Thysanoptera). Jour. Chem. Ecol. 11: 35–43.

Kjaer, A. 1960. Naturally derived isothiocyanates (mustard oils) and their parent glucosides. Fortshr. Chem. Org. Naturst. 18: 122–176.

Klingenburg, M. and J. Bucher. 1960. Biological oxidations. Annu. Rev. Biochem. 29: 669–708.

Kogan, M. 1976. The role of chemical factors in insect/plant interrelationships. Proc. XV Int. Cong. Entomol. Washington, D.C., 211–227.

Kogan, M. 1982. Plant resistance in pest management. Chapt. 4, in R.L. Metcalf and W. H. Luckman, eds. "Introduction to Insect Pest Management." 2nd ed., John Wiley & Sons, N.Y.

Kramer, E. 1978. Insect pheromones, pp. 205–229, in G.L. Hazelbauer, ed. "Taxis and Behavior." Chapman & Hall, London.

Kumar, N. and M.G. Motts. 1986. Volatile constituents of peony flowers. Phytochemistry 25: 250–253.

Labandeira, C.C., B.S. Beall and F.M. Huber. 1988. Early insect diversification: evidence of a lower Devonian bristletail from Quebec. Science 242: 913.

Ladd, T.L. Jr., B.R. Stinner and H.R. Krueger. 1983. Eugenol a new attractant for the northern corn rootworm (Coleoptera: Chrysomelidae). Jour. Econ. Entomol. 76: 1049–1051.

Lampman, R.L. and R.L. Metcalf. 1988. The comparative response of *Diabrotica* species (Coleoptera: Chrysomelidae) to volatile attractants. Environ. Entomol. 17: 644–648.

Lampman, R.L., R.L. Metcalf and J.F. Andersen. 1987. Semiochemical attractants of *Diabrotica undecimpunctata howardi* Barber, southern corn rootworm, and *Diabrotica virgifera virgifera* LeConte, the western corn rootworm (Coleoptera: Chrysomelidae). Jour. Chem. Ecol. 13: 959–975.

Lewis, P.A., R.L. Lampman and R.L. Metcalf. 1990. Kairomonal attractants for *Acalymma vittatum* (Coleoptera; Chrysomelidae). Environ. Entomol. 19: 8–14.

Ma, W.C. and J.H. Visser. 1978. Single unit analysis of odour quality coding by the olfactory antennal system of the Colorado beetle. Entomol. Exp. Appl. 24: 520–523.

MacLeod, A.J., N.M. Pieris and V. Gill. 1981. Volatile constituents of sugar beet leaves. Phytochemistry 20: 2292–2295.

MacLeod, A.J. and N.G. deTronconis. 1982a. Volatile flavor components of guava. Phytochemistry 21: 1339–1342.

MacLeod, A.J. and N.G. deTronconis. 1982b. Volatile flavor components of mango fruit. Phytochemistry 21: 2523–2526.

MacLeod, A.J. and M.L. Nussbaum. 1977. The effects of different horticultural practices on the chemical flavor components of some cabbage cultivars. Phytochemistry 16: 861–865.

Mathis, C. and G. Ourisson. 1963. Étude chimio-taxonomique du genre *Hypericum*. I. Répartition de l' hypericine. Phytochemistry 2: 157–171.

Matsumoto, Y.A. 1962. A dual effect of coumarin, olfactory attraction and feeding inhibition of the vegetable weevil in relation to the uneatability of sweet clover leaves. Japan. J. Appl. Entomol. Zool. 6: 141–149.

Matsumoto, Y.A. and S. Sugiyama. 1960. Studies on the host plant determination of the leaf feeding insects. V. Attraction of leaf alcohol and some aliphatic alcohols to the adult and larvae of the vegetable weevil. Ber. Ohara Inst. Landwirtsch Biol. 11: 359–364.

Matsumoto, Y.A. and A.J. Thorsteinson. 1968. Effect of organic sulfur compounds on oviposition in onion maggot *Hylemyia antiqua* Meig. Appl. Entomol. Zool. 3: 5–12.

Meeuse, B.J.D. 1978. The physiology of some sapromyophilous flowers, pp. 97–104 in A.J. Richards, ed. "Pollination of Flowers by Insects." Academic Press, N.Y.

Metcalf, R.L. 1985. Plant kairomones and insect pest control. Bull. Ill. Nat. Hist. Sur. 33: 175–198.

Metcalf, R.L. 1986. Coevolutionary adaptations of rootworm beetles (Coleoptera: Chrysomelidae) to cucurbitacins. J. Chem. Ecol. 12: 1109–1124.

Metcalf, R.L. 1987. Plant volatiles as insect attractants. CRC Critical Rev. Plant Sci. 5: 251–301.

Metcalf, R.L. 1990. Chemical ecology of Dacinae fruit flies (Diptera: Tephritidae). Ann. Entomol. Soc. Amer. 85: 1017–1030.

Metcalf, R.L. and R.L. Lampman. 1989a. Chemical ecology of Diabroticites and Cucurbitaceae. Experientia 45: 240–247.

Metcalf, R.L. and R.L. Lampman. 1989b. Estragole analogues as attractants for Diabroticite species (Coleoptera: Chrysomelidae) corn rootworms. J. Econ. Entomol. 82: 123–129.

Metcalf, R.L. and R.L. Lampman. 1989c. Cinnamyl alcohol and analogs as attractants for corn rootworms (Coleoptera: Chrysomelidae). J. Econ. Entomol. 82: 1620–1625.

Metcalf, R.L., E.R. Metcalf and W.C. Mitchell. 1979. Evolution of the olfactory receptor in oriental fruitfly *Dacus dorsalis*. Proc. Natl. Acad. Sci. U.S.A. 76: 1561–1565.

Meyer, H.J. and D.M. Norris. 1967. Vanillin and syringaldehyde as attractants for *Scolytus multistriatus* (Coleoptera: Scolytidae). Ann. Entomol. Soc. Amer. 60: 858–859.

Millar, J.G., Z. Cheng-hua, G.N. Lanier, D.P. O'Callagan, M. Griggs, J.R. West and R.M. Silverstein. 1986. Components of moribund American elm trees as attractants to elm bark beetles, *Hylurgopinus rufipes* and *Scolytus multistriatus*. J. Chem. Ecol. 12: 583–608.

Miller, J.R., M.O. Harris and J.A. Brezrak. 1984. Search for potent attractants of onion flies. Jour. Chem. Ecol. 10: 1477–1488.

Minyard, J.P., D.D. Hardee, R.C. Gueldner, A.C. Thompson, G. Wiygul and P.A. Hedin. 1969. Constituents of the cotton bud. Compounds attractive to the boll weevil. Jour. Agric. Food Chem. 17: 1093–1097.

Morgan, A.D. and S.C. Lyon. 1928. Notes on amyl salicylate as an attractant to the tobacco hornworm moth. Jour. Econ. Entomol. 21: 189:191.

Mustaparta, H. 1990. Chemical information processing in the olfactory system of insects. Physiol. Rev. 70: 199–214.

Nishida, R. and H. Fukami. 1989. Oviposition stimulants of an Aristolochiae-feeding swallowtail butterfly, *Atrophaneura alcinous*. Jour. Chem. Ecol. 15: 2565–2575.

Nishida, R., T. Ohsugi, S. Kokubo and H. Fukami. 1987. Oviposition stimulants of a citrus feeding swallowtail butterfly *Papilio xuthus* L. Experientia 43: 342–344.

Norlander, G.H., H. Erdmann, U. Jacobsen and K. Siodin. 1986. Orientation of the pine weevil *Hylobius abietis* to underground sources of host volatiles. Entomol. Exp. Appl. 41: 91–100.

Nyar, J.K. and G. Fraenkel. 1962. The chemical basis of host selection in the Mexican bean beetle *Epilachna varivestis*. Ann. Entomol. Soc. Amer. 56: 174–178.

Pereya, P.C. and M.D. Bowers. 1988. Iridoid glycosides as oviposition stimulants for the buckeye butterfly *Junonia coenia*, nymphalid butterfly. Jour. Chem. Ecol. 14: 917–928.

Price, P.N. 1984. "Insect Ecology." 2nd. ed. John Wiley & Sons, Inc. N.Y.

Priesner, E. 1973. Reaktionen von Riechrezeptoren männlicher Solitarbienen auf inhalstoffe von Ophyrs-Bluten. Zoon. Suppl. 1: 43–55.

Rees, C.J.C. 1969. Chemoreceptor specificity associated with a choice of feeding sites by the beetle *Chrysolina brunvicensis* on its food plant, *Hypericum hirsutum*. Entom. Exp. Appl. 12: 565–583.

Rice, E.L. 1974. "Allelopathy". Academic Press, N.Y.

Riek, E.F. 1970. Fossil history, pp. 168–186 in "Insects of Australia". Melbourne University Press, Melbourne, Australia.

Ripley, L.B. and G.A. Hepburn. 1935. Olfactory attractants for male fruit flies. Entomol. Mem. Dept. Agr. S. Africa. 9: 3–17.

Rodriguez, E., P.L. Healey and I. Mehta 1984. Eds. "Biology and Chemistry of Plant Trichomes." Plenum Press, N.Y.

Schneider, D. 1968. Insect antennae. Annu. Rev. Entomol. 9: 103–122.

Schneider, D. 1955. Mikro-Electroden registrieren die elektrischen Impulse enzelner Sinnesnervekellen der Schmetterlingsantenne. Ind. Elktron. (Hamburg) 3: 3–7.

Schneider, D. 1957a. Elektrophysiologische Untersuchungen von Chemo und Mechanorezeptoren der Antenne des Seidenspinners *Bombyx mori* L. Z. Vgl. Physiol. 40: 8–41.

Schneider, D. 1957b. Electrophysiological investigations on the antennal receptors of the silk moth during chemical and mechanical stimulation. Experientia (Basel) 13: 89–91.

Schneider, D., B.C. Block, J. Boeckh and E. Priesner 1967. Die reaction der Mannlichen Seidenspinner auf Bombykol und seine Electroantennogramen und Verhalten. Z. Vgl. Physiol. 54: 192–209.

Schoonhoven, L.M. 1972. Secondary plant substances and insects. Rec. Adv. Phytochem. 5: 197–224.

Schoonhoven, L.M. 1985. Insects in a chemical world, pp. 1–21 in E.D. Morgan and N.B. Mandava, eds. "Handbook of Natural Pesticides." Vol. VI. CRC Press, Boca Raton, FL.

Schwartz, P.H. Jr., D.W. Hamilton and B.G. Townshend. 1970. Mixtures of compounds as lures for the Japanese beetle. Jour. Econ. Entomol. 63: 41–43.

Slifer, E.H. and S.S. Sekhon. 1961. Fine studies of the sense organs on the antennal flagellum of the honeybee *Apis mellifera*. J. Morphol. 109: 351–380.

Smart, J. and N.F. Hughes. 1973. The insect and plant progressive paleological integration, pp. 143–155 in H.F. Van Emden, ed. "Insect/Plant Interrelationships", Entomol. Soc. London.

Smilanick, J.M., L.E. Ehler and M.C. Birch. 1975. Attraction of *Carpophilus* spp. (Coleoptera: Nitidulidae) to volatile compounds present in figs. J. Chem. Ecol. 4: 701–707.

Son, K-c, R.F. Severson, R.F. Arrendale and S.J. Kays. 1990. Isolation and characterization of pentacyclic triterpene ovipositional stimulant for the sweet potato weevil from *Ipomoea batatas* (L.) Lam. J. Agr. Food Chem. 38: 134–137.

Stadler, E. 1984. Contact chemoreception, pp. 1–35 in W.J. Bell and R.T. Carde eds. "Chemical Ecology of Insects." Chapman & Hall, London.

Thomas, H. and G. Hertell. 1969. Responses of the Pales weevil to natural and synthetic host attractants. J. Econ. Entomol. 62: 383–386.

Van der Pers, J.N.C. 1981. Comparison of electroantennogram response spectra to plant volatiles in seven species of *Yponmeuta* and in the tortricid *Adoxophyes orana*. Entomol. Exp. Appl. 30: 181–192.

Vernon, R.S., G.J.R. Jubb, J.H. Borden, H.D. Pierce Jr. and A.C. Onschlagen. 1981. Attraction of *Hylemya antiqua* (Meigen) (Diptera: Anthomyiidae) in the field to host produced oviposition stimulants and their non-host analogues. Can. J. Zool. 5(; 872–881.

Visser, J.H. 1979. Electroantennographic responses of the Colorado beetle, *Leptinotarsa decemlineata* to plant volatiles. Entomol. Exp. Appl. 25: 86–97.

Visser, J.H. 1983. Differential sensory perceptions of plant compounds by insects, pp. 215–220 in P.A. Hedin, ed. "Plant Resistance to Insects". Amer. Chem. Soc. Symp. Ser. 208, Washington, D.C.

Visser, J.H. 1986. Host odor perception in phytophagous insects. Annu. Rev. Entomol. 31: 121–144.

Visser, J.H. and D.A. Ave. 1978. General green leaf volatiles in the olfactory orientation of the Colorado potato beetle, *Leptinotarsa decemlineata*. Entomol. Exp. Appl. 24: 438–449.

Visser, J.H. and J.R. Nielsen. 1977. Specificity in the olfactory orientation of the Colorado beetle, *Leptinotarsa decemlineata*. Entomol. Exp. Appl. 21: 14–22.

Vogel, S. 1966. Scent organs of orchid flowers and their relation to insect pollination, pp. 253–259 in L.R. DeGarmo, ed. "Proc. 5th World Orchid Conf.", Long Beach, CA.

Von Frisch, K. 1967. "The Dance Language and Orientation of Bees." Harvard Univ. Press, Cambridge, MA.

Wearing, C.H. and K.F.N. Hutchins. 1973. Alpha-farnesene a naturally occurring oviposition stimulant for the codling moth, *Laspeyresia pomonella*. J. Insect Physiol. 19: 1251–1256.

Wenner, A.M. 1974. Information transfer in honeybees. A population approach, pp. 113–169, in L. Kramer, P. Pilner and T. Alloway, eds. "Nonverbal Communication." Plenum Press, NY.

Wiebes, J.T. 1974. Coevolution of figs and their insect pollinators. Annu. Rev. Ecol. Syst. 10: 1–10.

Wilde, J. de. 1957. Vergetten Hoofdstukken Uit de Phytopharmacie. Ghent Landbouw Hogeschool. Meded. 22: 335–347.

Williams, M. 1986. The receptor from concept to function. Chapt. 21 in R.W. Egan, ed. Ann. Rept. Medicinal Chem. Academic Press, N.Y.

Williams, N.H. 1981. The biology of orchids and Euglossini bees, pp. 119–171 in J. Arditti, ed. "Orchid Biology, Reviews and Perspectives II." Cornell Univ. Press, Ithaca, NY.

Williams, N.H. 1983. Floral fragrances as cues in animal behavior. pp. 50–71, in G.E. Jones and R.J. Little eds. "Handbook of Experimental Pollination Biology." Scientific and Academic Editors, N.Y.

Williams, N.H. and W.M. Whitten. 1983. Orchid floral fragrances and male Euglossine bees: methods and advances in the last sesquidecade. Biol. Bull. (Woods Hole) 163: 355–395.

Wilson, E.O. 1970. Chemical communication within animal species, pp. 133–155, in E. Sondheimer and J.B. Simeone, eds. "Chemical Ecology." Academic Press, N.Y.

Yamaguchi, K. and T. Shibamoto. 1980. Volatile constituents of the chestnut flower. Jour. Agr. Food Chem. 28: 82–84.

Yaro, N.D., J.D. Krysan and T.L. Ladd, Jr. 1987. *Diabrotica cristata* (Coleoptera: Chrysomelidae) attraction to eugenol and related compounds compared with *D. barberi* and *D. virgifera virgifera*. Environ. Entomol. 16: 126–128.

2

VOLATILE KAIROMONES AS LURES FOR INSECTS

I. INTRODUCTION

A number of volatile plant kairomones have practical use as lures to attract insects to traps for monitoring populations or to baits for control. This technology began with Howlett's (1915) discovery that methyl eugenol from lemon grass oil *Cymbopogon nardus* was highly attractive to the male fruit flies *Dacus diversus* and *D. zonatus* (see Chapter 5). It should be noted that this discovery marked the beginning of the modern chemical ecology of insects (Metcalf 1990), and that this first chemical characterization of a plant kairomone for insect control antedated the discovery of the first insect pheromone bombykol ((*E*,*Z*)10,12-dodecadienol), the sex pheromone of the silkworm moth *Bombyx mori* (Butenandt et al. 1959), by more than 40 years. Geraniol, a volatile kairomone for the Japanese beetle *Popillia japonica*, was patented by Smith et al. (1926) and mixtures of the plant kairomones geraniol, eugenol, and phenethyl alcohol esters are widely used today to control this pest (Chapter 3). In 1935, Ripley & Hepburn showed that terpineol acetate present in a variety of essential oils was a specific attractant for the Mediterranean fruit fly *Ceratitis capitata* and the Natal fruit fly *C. rosae* (Chapter 5).

Despite these promising beginnings, research on and development of plant kairomone attractants for insect pest control were neglected during the period after World War II because of the euphoria over the development of DDT and other synthetic insecticides (Metcalf 1980), and again during the exciting developments of insect sex pheromones during the 1960's and 1970's.

II. ACTIVE ODOR SPACE

Bossert & Wilson (1963) in a seminal paper introduced the concept of "active space", i.e. the three dimensional zone downwind from the point of emission of a volatile semiochemical where the molecular concentration is above the threshold required for activation of the insect behavioral response. This zone is the signal from the kairomone releaser, either from a natural plant source or from a baited kairomone trap. Within the dimensions of the active space, the perceiving insect may respond, following receptor activation, by a stereotyped behavioral response.

The ratio of the emission rate of the kairomone Q in molecules sec^{-1} to the threshold concentration for the insect response K in molecules ml^{-1} is a critical factor in volatile semiochemical communication. The larger the Q/K ratio, the greater the maximum distance that the signal is effective and the slower the signal fadeout. Therefore, the shape and volume of the active space thus portrayed is of fundamental importance in the behavioral pattern of the insect attracted, and in understanding the manner in which volatile kairomones can be used for monitoring and controlling insect pests.

A. Quantitative Aspects of Active Space

Very few examples of chemical communication occur in still air and on planar unobstructed surfaces. Dispersion of semiochemicals in nature involves distribution downwind over uneven terrain and with eddy currents induced by vegetation. Wright (1958) was the first to model the dispersal of odorants using the Sutton (1953) gas diffusion model for downwind dispersion from a single point source of continuous emission:

$$U(x,y,z) = \frac{2\,Q}{\pi C_y C_z u x^{2-n}}\, e^{-x^{n-2}} \left(\frac{y^2}{C_{y2}} + \frac{z^2}{C_{z2}}\right) \tag{1}$$

where U is the concentration at the downwind coordinates x,y,z with the source emitting at a constant rate Q (molecules per second) in a steady wind velocity u (cm sec^{-1}); n is an index that varies between 0 and 1 with the vertical wind profile; and C_y and C_z are the horizontal and vertical dispersion coefficients that vary with the wind velocity u and the roughness of the terrain. Sutton (1953) suggested that dispersion in winds of 100 to 500 cm sec^{-1} (2.2 to 11 mph) and over relatively level terrain, can be modelled satisfactorily using typical values of $n = 0.25$, $C_y = 0.4$ $cm^{0.125}$ and $C_z = 0.2$ $cm^{0.125}$.

To approximate the maximum dimensions of the active odor space in terms of Q (emission rate of the semiochemical in molecules sec^{-1}) and K (behavioral threshold in molecules ml^{-1}) Bossert & Wilson (1963) modified the Sutton equation (1) to incorporate the Q/K ratio:

$$K = \frac{2\,Q}{\pi C_y C_z u x^{2-n}}\, e^{-x^{n-2}} \left(\frac{y^2}{C_{y2}} + \frac{z^2}{C_{z2}}\right) \tag{2}$$

From this equation, the maximum distance downwind in cm above the threshold K is (Bossert & Wilson 1963):

$$X_{max} = \left(\frac{2\,Q}{K\pi C_y\,C_z u}\right)^{1/2-n} \tag{3}$$

The maximum height

$$Y_{max} = C_y\sqrt{\frac{2\,Q}{K\pi\,C_y\,C_z\,ue}} \tag{4}$$

and the maximum width

$$Z_{max} = C_z\sqrt{\frac{2\,Q}{K\pi\,C_y\,C_z\,ue}} \tag{5}$$

Equation (3) can be rearranged to:

$$Q/K = 1/2\,C_y\,C_z\,uX_{max}{}^{2-n} \tag{6}$$

and by inserting Sutton's typical values for $C_y = 0.4\ cm^{0.125}$, $C_z = 0.2^{0.125}$ and $n = 0.25$; equation (6) simplified to (Bossert & Wilson 1963):

$$Q/K = 0.125\ u\ X_{max}{}^{7/4} \tag{7}$$

Equations (6) and (7) are useful for the approximation of the Q/K ratio and for the approximation of K, the behavioral threshold, which is difficult to determine directly.

The modified Sutton equation (3) has been used to estimate maximum distances for pheromone communication between males and females of the cabbage looper, *Trichoplusia ni* (Sower et al. 1971, 1973), the oriental fruit moth *Laspeyresia molesta* (Baker & Roelofs 1981), and the cotton leafworm *Spodoptera litura* (Nakamura & Kawasaki 1977). The simulations were in reasonable agreement with observations on the behavior of responding male moths.

The Sutton equation incorporates average dispersion coefficients and these apply only to condition of neutral atmospheric stability. Equation (1) can be applied to unstable atmospheric conditions by using different values for n and for the dispersion coefficients, C_y and C_z.

Theoretical objections have been directed at the use of the Sutton equation, particularly under conditions of atmospheric instability and in the presence of obstructive vegetation. Fares et al. (1980) used Gaussian distributions around both lateral and vertical components downwind from a ground level source to develop equation (8) which they used in analogy with equation (2) of Bossert & Wilson (1963):

$$C(x,0,0;0) = \frac{Q(1 + \alpha)}{2\pi\sigma_y(x)\sigma_z(x)u} \tag{8}$$

where C = concentration at point x downwind in mol m^{-3}, Q = average rate of semiochemical emission in mol sec^{-1}, $\sigma_y(x)$ and $\sigma_z(x)$ are the lateral and vertical diffusivities of the plume concentration in m, α is a reflectance coefficient that may vary from 0–1, and u is the wind speed in cm sec^{-1}. When α is set at 1 to relate to total reflectance and the threshold for response K in mol cm^{-3} is used to determine the boundaries of the active space generated, equation (8) can be written as:

$$K = \frac{Q}{2\pi\sigma_y(x)\sigma_z(x)u}\exp\left[-\frac{1}{2}\left(\frac{y}{\sigma_y(x)}\right)^2\right] \tag{9}$$

In practice, the use of equation (9) is complicated because the downwind distance (x) does not appear explicitly in the equation and the dispersion coefficients $\sigma_y(x)$ and $\sigma_z(x)$ depend upon atmospheric stability and vary with the downwind distance so that they must be determined empirically.

To simulate the maximum horizontal area of short, ground level odorant plumes, Stanley et al. (1985) employed the modified equation (10):

$$A_R = 2\text{ Fa } \exp^{[-1/2]}\left[\frac{b + d}{b}\right]^{1/2}\left(\frac{R}{2\pi ac}\right)^{\frac{(b + 1)}{(b + d)}} \tag{10}$$

where A_R = area of plume and R in m^2 is the effective communication factor adjusted for wind speed, i.e. R = Q/Ku. By setting R = 1, the area of the plume can be calculated by the power function:

$$A_R = A_1 R^{\beta} \tag{11}$$

where $\beta = (b + 1)/(b + d)$ (b and d are tabular horizontal and vertical dispersion constants).

B. Behavioral Threshold K

Direct measurement of the insect behavioral threshold to odorants is difficult and only a few determinations have been made using combinations of gas chromatography and electroantennography. Boeckh et al.

(1965) determined that the threshold value K for the response of the grasshopper *Locusta migratoria* to the green volatile hexenol was 1×10^8 mol cm^{-3}. In a similar study, Visser (1979) determined that the Colorado potato beetle, *Leptinotarsa decemlineata* responded to *trans*-2-hexen-1-ol at 1.2×10^8 and to *cis*-3-hexen-1-ol at 1.2×10^{11} mol cm^{-3}. Finch (1980) has emphasized that the relationship between the electrophysiological threshold and the behavioral threshold is uncertain and that the latter is likely to be at least 0.1 that of the former.

The modified Sutton equation (3) affords a convenient means of incorporating values for Q and K into a practical estimate of the dimensions of active space from single kairomone lures. This equation provides for relatively simple calculations of the dimensions of "active space" that are realistic in terms of actual field observations. The rearranged equation (7) provides the only practical way to estimate the insect behavioral threshold response K, which is critical to the estimation of active space. The Sutton equation is most useful in comparative evaluations of lures, e.g. kairomones vs. parakairomones, release rates from various formulations, and the effects of changes in wind velocity and of occluded terrains.

1. Effective Distance X_{max} for Kairomone Attractivity

There is little data in the literature about the maximum distance that plant kairomones can attract insects. Steiner (1952) reported that a film of 5 g of methyl eugenol attracted male oriental fruit flies, *Dacus dorsalis* from a distance of 0.5 to 1 mile, against a wind of 8 mph (356 cm sec^{-1}). The calculated release rate of methyl eugenol was 2.5×10^{17} mol sec^{-1} (Table 2.1). Using Equation (3) and assuming a range of values for K, the estimated values for X_{max} were:

K (assumed, mol cm^{-3})	X_{max} (estimated, m)
1×10^5	3084
1×10^6	1024
1×10^7	275

These data suggest that the most probable value of K for *D. dorsalis* is approximately 1×10^6 mol cm^{-3}.

Finch (1980) used Equation (3) to estimate the distance of attraction (X_{max}) for the cabbage maggot fly *Delia brassicae* to a single brassicaceous plant liberating allylisothiocyanate at 7 μl per day (4.8×10^{13} mol sec^{-1}). With the assumption that $K = 6 \times 10^5$ mol cm^{-3}, the calculated values were: wind 45 cm sec^{-1}, $X_{max} = 63.8$ m and wind 450 cm sec^{-1}, $X_{max} =$

Table 2.1. Release Rates for Kairomone Lures at 24°C*

lure	mol. wt.	bp(°C)	release rate for 20 mg (mg hr^{-1})	(mol. sec^{-1})
	lures for Dacinae			
methyl eugenol	178	254	1.1	1.0×10^{15}
raspberry ketone	164	340	0.00084	8.5×10^{11}
raspberry ketone, formyl ester	192		0.029	2.6×10^{13}
raspberry ketone, acetyl ester (cue-lure)	206	345	0.016	1.3×10^{13}
raspberry ketone, propanyl ester	220		0.01	7.6×10^{12}
raspberry ketone, butanyl ester	234		0.0066	4.8×10^{12}
	lures for Diabroticites			
cinnamaldehyde	132	248	1.1	1.4×10^{15}
cinnamyl alcohol	134	250	0.21	2.5×10^{14}
cinnamonitrile	129	254	1.1	1.4×10^{15}
estragole	148	215	10.9	1.2×10^{16}
eugenol	164	254	1.25	1.3×10^{15}
indole	117	253	1.0	1.4×10^{15}
β-ionone	192	265	0.50	4.4×10^{14}
4-methoxycinnamaldehyde	162		0.015	1.6×10^{13}
4-methoxycinnamonitrile	159		0.096	1.1×10^{14}
4-methoxyphenethanol	152	335	0.11	1.2×10^{14}
3-phenpropanol	136	235	0.82	1.1×10^{15}
1,2,4-trimethoxybenzene	168	247	0.19	1.9×10^{14}

* Data from Metcalf & Lampman (1991), Metcalf & Mitchell (1990).

17.1 m. These calculated values were compared with a field observation that *D. brassicae* was attracted to brassicae over a distance of at least 15 m (Finch 1980).

Movements of adult western corn rootworms, *Diabrotica virgifera virgifera*, from corn upwind to lures placed in an open alfalfa field were studied by Lampman et al. (1991). Replicated cylindrical sticky traps baited with 100 mg of 4-methoxycinnamaldehyde were placed at distances of 10, 30, and 100 m from the edge of the infested corn field. The ratios of the capture rates of *D. v. virgifera* for attractant/untreated control were 10 m (59 times), 30 m (30 times), and 100 m (20 times). These values were significant at P = 0.001. Values of X_{max} were calculated from Equation (3) with Q for 4-methoxycinnamaldehyde = 8.0×10^{13} mol sec^{-1}.

K (assumed, mol cm^{-3})	X_{max} (estimated, m)
1×10^{4}	294
1×10^{5}	79
1×10^{6}	21
1×10^{7}	6

These results suggest that K for *D. v. virgifera* exposed to 4-methoxy-cinnamaldehyde is approximately 1×10^5 mol cm^{-3}.

C. Determination of the Release Rate Q

The release rate of a kairomone is readily determined by measuring the loss of weight or of volume of a standard sample of the semiochemical with time. Wilson et al. (1969) measured the change in volume of liquid semiochemicals in capillary tubes using an ocular micrometer. McGovern et al. (1966) used filter paper discs impregnated with approximately 50 mg of lure suspended in a hood. The papers were weighed at intervals to determine weight loss. The relative weight losses of the several attractants of interest here, at 27°C were:

	Volatilization	
Lure	(mg min^{-1})	(mg day^{-1})
methyl eugenol	0.032	46.0
cue-lure	0.00043	0.62
medlure	0.034	49.0
trimedlure	0.066	95.0
siglure	0.25	346.0

In our laboratory (Metcalf & Lampman 1991) kairomone release rates are determined by weighing 20 mg of lure into 30 mm aluminium planchets. The test chemical was distributed as evenly as possible over the bottom of the planchet by adding 0.1 to 0.2 ml of acetone. After the acetone evaporated the planchets were weighed and placed in a ventilated hood at a constant air velocity of 75 cm sec^{-1} (150 ft min^{-1} or 1.7 miles $hour^{-1}$) at 23–25°C. The planchets were weighed at appropriate intervals and the loss in weight was determined as the average of four replications. Rates of volatilization as shown in Table 2.1 were calculated from the initial approximately straight line portion of the curve of weight loss vs. time. Release rates of kairomone lures from thin films are directly proportional to the amount (first order) and kairomone trap release can be approximated using appropriate multiples of the values in Table 2.1.

The relative chemosensory activity of the kairomone lures in field studies is a function of the relative release rate and of the limit of response (LR) (Figure 4.11) (Metcalf & Lampman 1991). For example, for *D. v. virgifera*, cinnamaldehyde has a release rate ≈ 83 times greater than 4-methoxycinnamaldehyde and the latter is ≈ 33 times more effective in

LR. Therefore the relative chemosensory activity of 4-methoxycinnamaldehyde is $83 \times 33 = 2750$ times greater than cinnamaldehyde.

III. PARAKAIROMONES

Parakairomones are synthetic analogues of naturally occurring semiochemicals that produce behavioral responses in target insects that are similar to those produced by exposure to naturally occurring kairomones. There are numerous examples of parakairomones that are discussed in this book, and some of these have become important for monitoring insect populations and as lures for toxic baits. The lures for the Mediterranean fruit fly, *Ceratitis capitata*, siglure, medlure, and trimedlure are parakairomones of the natural product α-copaene (Chapter 5). The lures for the melon fly *Dacus cucurbitae*, anisyl acetone and cue-lure, are parakairomones related to the natural product raspberry ketone (Chapter 5). The lures for the oriental fruit fly *Dacus dorsalis*, 3,4-dimethoxypropylbenzene, 3,4-dimethoxybenzyl methyl ether and 3,4-dimethoxyphenyl ethyl ether are parakairomones of methyl eugenol (Mitchell et al. 1985) (Chapter 5).

For the *Diabrotica* rootworm beetles, cinnamonitrile is a parakairomone of cinnamaldehyde for the southern corn rootworm, *D. u. howardi*; 4-methoxycinnamonitrile is a parakairomone of 4-methoxycinnamaldehyde for the western corn rootworm, *D. v. virgifera*; and 3-phenylpropanol is a parakairomone for cinnamyl alcohol for the northern corn rootworm, *D. barberi*. (Chapter 4).

Typically, these parakairomones are structurally optimized kairomones and may have simpler chemical structures which are much cheaper to produce, e.g. trimedlure vs. α-copaene. Parakairomones may have more favorable chemical properties in regard to volatility and persistence. Cue-lure with a release rate of about 17 times greater than raspberry ketone is a more effective long range attractant for *D. cucurbitae* although it is less persistent. Parakairomones may have more favorable toxicological properties, e.g. 3,4-dimethoxy-1-propylbenzene as a substitute for methyl eugenol, which is weakly carcinogenic.

A. Parakairomones as Bioisosteres

As we have seen (Chapter 1) the receptor is a membrane recognition site that is capable of recognizing subtle differences in the chemical composition of ligands (Williams 1986). Kairomone receptors are key factors in the insects' recognition of semiochemicals and hence in their response to lures which are ligands that act as agonists in depolarizing chemosensory receptors. Bioisosterism is the key factor in designing parakairomones. Bioisosteres are defined as molecules which have chemical and physical similarities producing broadly similar biological properties (Lapinski 1986). The study of bioisosterism serves to delineate the structural requirements of the agonist necessary to interact with the receptor membrane and to depolarize the receptor. Conversely, molecules designed as bioisosteres are in effect specific molecular probes for determining the dimensions of the receptor. Properties that are important in receptor interactions include size, shape, electron distribution, chemical reactivity, hydrogen bonding, pK_a and lipid/H_2O partitioning (Lapinski 1986).

B. Limits of Response to Kairomones and Parakairomones

The limit of response (LR) is defined as the least quantity of semiochemical which, under standard conditions, attracts the test insect to treated baits where they display the characteristic behavioral response (Metcalf et al. 1983). Obviously, the better the fit of the parakairomone to macromolecular receptor site, the greater the receptor depolarization and the lower the LR value. Therefore, the determination of the LR provides a quantitative measure of the effectiveness of the lure. Examples of the value of LR determinations are shown in Figures 4.6 and 4.7.

IV. TRAPS FOR KAIROMONE LURES

Innovation is essential in the design of kairomone baited traps for monitoring and controlling insect populations. Thousands to hundreds of thousands of these traps may be deployed for monitoring Tephritidae fruit flies (Chapter 5) and for controlling Japanese beetles (Chapter 3). Simplicity, durability, and minimal cost are key factors. Representative kairomone trap designs are shown in Figure 2.1.

Trap design is dependent upon the chemotactic behavior of the insect responding to the kairomone lure. For example, Tephritidae fruit flies readily enter the interiors of traps in response to such powerful kairomone attractants as methyl eugenol and cue-lure. Early investigators used the all glass, invaginated McPhail trap with various liquid bait formulations that effectively drowned the flies. Steiner (1957) enticed the flies through a restricted opening of a plexiglass trap to a wick treated with methyl eugenol and naled insecticide (Figure 2.1). More recent developments were based on the use of sticky traps coated with insect adhesive (Harris et al. 1971). These investigators compared the responses of three Te-

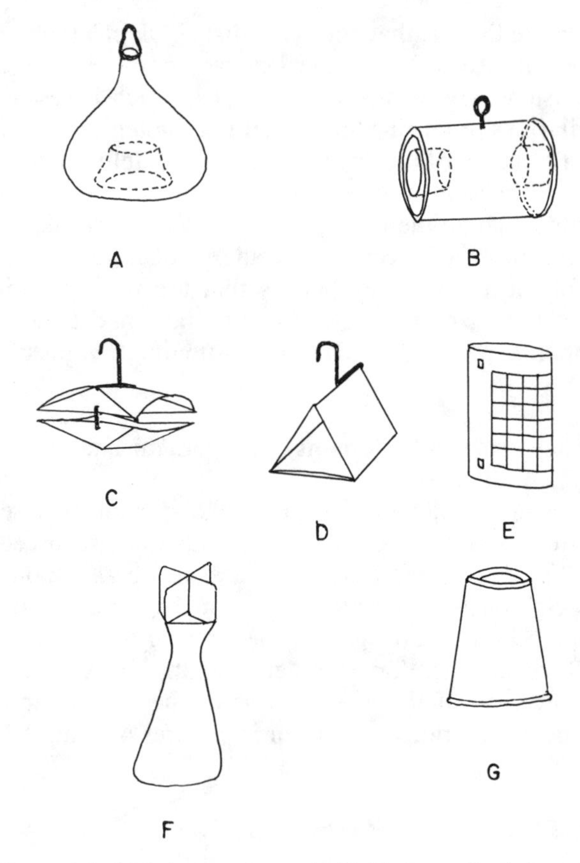

Figure 2.1. Traps used with kairomone lures: A—McPhail trap, B—Steiner trap, C—winged sticky trap, D—delta trap or Jackson trap, E—rectangular "apple maggot" trap, F—Japanese beetle trap, and G—yellow cup cylindrical trap.

phritidae spp. *Dacus dorsalis* to methyl eugenol, *D. cucurbitae* to cue-lure, and *Ceratitis capitata* to trimedlure in Steiner traps, rectangular traps, and triangular traps. The performances of several trap designs were rated by a trap catch index. The indices for the medfly were: Steiner 100, rectangular 120, and triangular 90, for the oriental fruit fly: Steiner 100, rectangular 20, and triangular 90, and for the melon fly: Steiner 100, rectangular 120, and triangular 160. It was concluded that the triangular or delta trap (Figure 2.1) at a cost of $0.05–0.10 was as efficient as the Steiner trap costing $1.00–3.00. This delta design has become very widely used internationally as the "Jackson trap" and features a removable sticky panel that is inserted into the bottom of the trap. Lures are absorbed on small sections of dental wicking held in the center of the trap.

In contrast to the Tephritidae fruit flies, Diabroticite beetles will not readily enter enclosed or folded traps. These beetles, however, are efficiently captured by flat or cylindrical traps. Hein & Tollefson (1984) compared the catch of the northern *Diabrotica barberi* and western *D. v. virgifera* corn rootworm beetles by traps of several designs. When placed in corn at ear height, winged traps (Pherocon C) gave poor results, while yellow rectangular traps (Pherocon AM) and yellow cylindrical traps (0.95 l) coated with insect adhesive were effective in trapping both species in sex ratios approximating those present in the field. In our experiments (C.D. Reid, R.L. Lampman and R.L. Metcalf, unpublished), sticky traps baited with 30 mg of 4-methoxycinnamaldehyde had the following 1 day catches of *D. v. virgifera*: delta traps 2.7 ± 1.3a, winged traps (Pherocon C) 17.5 ± 12.1a, yellow rectangular traps (Pherocon AM) 84.5 ± 11.3c, cylindrical traps 58.0 ± 1.9b, and yellow cup traps (Solo) (Levine & Metcalf 1988) 113.5 ± 20.03d.

The peculiarities of kairomone traps for the Japanese beetle are discussed in Chapter 3, and for the apple maggot in Chapter 5.

A. Trap Color

Many insects are preferentially attracted to the yellow-green portion of the spectrum (500–600 nm) and this wavelength range is typical of mature plant foliage and fruits (Landolt et al. 1988). Therefore, glossy bright yellow coloration is an important part of kairomone trap design. The Japanese beetle responds to bright yellow reflectance (Fleming et al. 1940) (Chapter 3) and commercial traps baited with kairomone lures contain yellow plastic vanes (Klein 1981) (Figure 2.1).

Both sexes of three representative species of Tephritidae, the cherry fruit fly *Rhagoletis cerasi*, the Mediterranean fruit fly, *Ceratitis capitata*, and the olive fly *Dacus oleae* have very similar spectral sensitivities with broad major peaks at 485–500 nm (yellow-green) and secondary peaks at 365 nm (ultraviolet). All three species are attracted to yellow panels showing strong reflectance maxima at 500–520 nm (Agee et al. 1982). Yellow rectangular panels were found to be attractive to *Anestrepha*, *Ceratitis*, *Dacus*, and *Rhagoletis* (Landolt et al. 1988) and the olive fly responded to various colored panels: fluorescent yellow (218a) > yellow (100) > light green (30b) > orange (22b) > red (5c) = clear (5c) > gray (3c) (Procopy and Economopoulos 1971). As a result of these and other studies, sticky yellow wing traps are widely used in Europe for monitoring and controlling the cherry fruit fly and yellow sticky boards are used extensively in Israel to control the olive fly (Agee et al. 1982). In the United States, the sticky yellow apple maggot trap (Figure 2.1) is widely used. The papaya fruit fly *Toxotrypana curvicauda* showed maximum visual sensitivity to wavelengths of about 475 and 500 nm, and sticky green spheres together with a slow release formulation of the sex pheromone 2,6-methylvinyl pyrazine have been incorporated into an effective lure (Landolt et al. 1988).

The northern corn rootworm *Diabrotica barberi* and the western corn rootworm *D. v. virgifera* adults have identical spectral responses with a major broad peak in the 500–540 nm (yellow green) range and a secondary peak at 365 nm (ultraviolet). There were no sexual differences in response (Agee et al. 1983). Ball (1982) measured the responses of adult *D. v. virgifera* to filters passing various colors at equal intensity and found the order of preference to be yellow (520–700 nm) > red (637–700 nm) > clear (350–700 nm) > light blue (400–800 nm) > dark blue (608–625 nm). Ladd et al. (1984) used fiberboard rectangles (15 × 20 cm) coated with insect adhesive and painted with various commercial colors to attract *D. barberi* adults. The order of attractivity was: yellow (100) > blue (80) > clear (65) > green (64) > fluorescent green (61) > fluorescent yellow (59) > fluorescent orange (51) > red (50) = blue (50).

B. Baited vs. Unbaited Traps

Rectangular yellow traps baited with 2 g eugenol caught 2–3 times more *D. barberi* adults than unbaited traps (Ladd et al. 1984). Field comparisons in corn of the efficacy of cylindrical sticky traps baited with 100 mg of TIC attractant (equal parts of 1,2,4-trimethoxybenzene, indole, and *trans*-cinnamaldehyde) (Chapter 4) showed that the 1 day catches of the kairomone baited traps had mean values of 38.4 times for *D. v. virgifera*, 11.7 times for *D. barberi*, and 13.2 times for *D. u. howardi*, greater than the unbaited traps (Lampman et al. 1991).

V. KAIROMONE TRAP DISTRIBUTION

The two major uses for kairomone trapping of insects are: monitoring for new infestations and removal trapping for pest control, and these pose somewhat different requirements for the efficient distribution of baited traps. Such considerations have been explored in some detail for the various Tephritidae fruit flies (Chapter 5) and to a limited extent for the Japanese beetle (Chapter 3).

Monitoring for fruit fly infestations has been conducted very extensively in California, Florida, and Australia. In the area around the Los Angeles International Airport, Steiner traps baited with methyl eugenol were deployed at the rate of 40 traps over the first 1.5 km and 12 traps over the next 5 km. This grid of methyl eugenol traps detected incipient outbreaks of *Dacus dorsalis* on five occasions over the period of 1961–1971 at a cost of about $225,000 (Chambers et al. 1974).

Kairomone traps baited with cue-lure are less efficient because of its much lower release rate (about 0.011 times that of methyl eugenol, Table 2.1). Trapping for the Queensland fruit fly *D. tryoni* in the suburbs of Adelaide is conducted in a 400 m grid, and male flies have been caught in 14 of the 17 years from 1960–1976. A similar 400 m grid of trimedlure baited traps caught male *Ceratitis capitata* on five occasions (Drew et al. 1978).

In removal trapping, the density of traps must be high enough to prevent the pest population from reaching the economic threshold. Mathematical analyses based on the release and recapture of marked insects has been discussed by Wolf et al. (1971) and Hartstack et al. (1971). The important factors are the area trapped by the individual trap (active space) and the trapping efficiency of the trap. Considered together these relate to the trap density function. The proposed calculations lead to graphs of the probability of insect pest capture at various trap spacings. Cunningham (1981) explored the theoretical regression curves relating to the performance of methyl eugenol/malathion baits for *D. dorsalis* and cue-lure/naled baits for *D. cucurbitae.* Hyperbolic and asymptotic regression lines were developed from considerations of trap performances at varying placements and dosages.

REFERENCES

Agee, H.R., E. Boller, U. Remund, J.C. Davis and D.L. Chambers. 1982. Spectral sensitivities and visual attractant studies on the Mediterranean fruit fly, *Ceratitis capitata* (Wiedmann), olive fly *Dacus oleae* (Gmelin) and European cherry fruit fly *Rhagoletis cerasi* (L.) (Diptera: Tephritidae). Zeit. Ang. Entomol. 93: 403–412.

Agee, H.R., V.M. Keith and J.C. Davis. 1983. Comparative spectral sensitivity and some observations on vision related behavior of northern and western corn rootworm adults. J. Ga. Entomol. Soc. 18: 240–245.

Ball, H.T. 1982. Spectral response of adult western corn rootworm (Coleoptera: Chrysomelidae) to selected wave lengths. J. Econ. Entomol. 75: 932–933.

Baker, T.C. and W.L. Roelofs. 1981. Initiation and termination of oriental fruit moth male response to pheromone concentrations in the field. Environ. Entomol. 10: 211–218.

Boeckh, J., K.E. Kaisling and D. Schneider. 1965. Insect olfactory receptors. Cold Spring Harbor Symp. Quant. Biol. 30: 263–280.

Bossert, W.H. and E.O. Wilson. 1963. The analysis of olfactory communication among animals. J. Theoret. Biol. 5: 443–469.

Butenandt, A.R., R. Beckmann, D. Stamm and E. Hecker. 1959. Uber den Sexuallockstoff den Seidenspinners *Bombyx mori.* Reindarstellung und Konstitution. Z. Naturforsch. B 14: 283–284.

Chambers, D.L., R.T. Cunningham, R.W. Lichty and R.B. Thrailkill. 1974. Pest control by attractants: a case study demonstrating economy, specificity, and environmental acceptability. Bioscience 24: 150–152.

Cunningham, R.T. 1981. The 3-body problem analogy in mass trapping programs, pp. 95–102, in E.R. Mitchell, ed. "Management of Insect Pests with Semiochemicals.", Plenum Press, N.Y.

Drew, R.A.I., G.H.S. Hooper and M.A. Bateman. 1978. "Economic Fruit Flies of the South Pacific Region." Watson Ferguson, Brisbane, Australia.

Elkington, J.S. and R.T. Carde. 1984. pp. 73–91 in W.H. Bell and R.T. Carde, eds. "Chemical Ecology of Insects.", Sinauer, Sunderland, MA.

Fares, Y., R.J. H. Sharpe and C.E. Magnusan. 1980. Pheromone dispersion in forests. J. Theoret. Biol. 84: 335–359.

Finch, S. 1980. Chemical attraction of plant-feeding insects to plants. pp. 67–143 in T.H. Coaker, ed. "Applied Biology." Vol V., Academic Press, N.Y.

Fleming, W.E., E.D. Burgess and W.W. Maines. 1940. The use of traps against the Japanese beetle. U.S. Dept. Agr. Cir. 594, 11 pp., Washington, D.C.

Harris, E.T., S. Nakagawa and T. Urago. 1971. Sticky traps for detection and survey of three Tephritidae. Jour. Econ. Entomol. 64: 62–65.

Hartstack, A.W. jr., J.P. Hollingsworth, R.L. Ridgway and H.H. Hunt. 1971. Determination of trap spacing required to control insect populations. J. Econ. Entomol. 64: 1090–1100.

Hein, G. and J.J. Tollefson. 1984. Comparison of adult corn rootworm trapping techniques. Environ. Entomol. 13: 266–271.

Howlett, F.M. 1915. Chemical reactions of fruit flies. Bull. Entomol. Res. 6: 297–305.

Jones, O.T., J.C. Lisk, C. Longhurst and P.E. Howse. 1983. Development of a monitoring trap for the olive fly *Dacus oleae* (Gmelin) (Diptera: Tephritidae) using a component of its sex pheromone as a lure. Bull. Entomol. Res. 73: 751–755.

Klein, M.G. 1981. Mass trapping for suppression of Japanese beetle, pp. 183–190 in E.R. Mitchell ed., "Management of Insect Pests with Semiochemicals." Plenum Press, N.Y.

Ladd, T.L., B.R. Stinner and A.R. Krueger. 1934. Influence of color and height of eugenol baited sticky traps on attractiveness to northern corn rootworm beetles (Coleoptera: Chrysomelidae). Jour. Econ. Entomol. 77: 652–654.

Lampman, R.L., R.L. Metcalf, L. Deem-Dickson and C.R. Reid. 1991. Attraction of Diabroticites (Coleoptera; Chrysomelidae) to a multicomponent lure and implications for corn rootworm baits. Environ. Entomol. In Press.

Landolt, P.J., R.R. Heath, H.R. Agee, J.H. Tumlinson and C.O. Calkins. 1988. Sex pheromone-based trapping system for papaya fruit fly (Diptera: Tephritidae). Jour. Econ. Entomol. 81: 1163–1169.

Lapinski, C.A. 1986. Bioisosteres in drug design, pp. 283–291 in D.M. Bailey ed. Ann. Rept. Med. Chem.

McGovern, T.P., M. Beroza, K. Ohinata and L.F. Steiner. 1966. Volatility and attractiveness to the Mediterranean fruit fly of trimedlure and its isomers and a comparison of its volatility with that of seven other insect attractants. J. Econ. Entomol. 59: 1450–1455.

Metcalf, R.L. 1980. Changing role of insecticides in crop protection. Annu. Rev. Entomol. 25: 219–256.

Metcalf, R.L. 1990. Chemical ecology of Dacinae fruit flies (Diptera: Tephritidae). Ann. Entomol. Soc. Amer. 83: 1017–1030.

Metcalf, R.L. and R.L. Lampman. 1991. Evolution of diabroticite rootworm beetle (Chrysomelidae) receptors for *Cucurbita* blossom volatiles. Proc. Nat. Acad. Sci. (USA) 88: 1869–1872.

Metcalf, R.L. and W.C. Mitchell. 1990. Design of more effective lures for the melon fly *Dacus cucurbitae*. Rept. to Calif. Dept. Food Agr.

Metcalf, R.L., W.C. Mitchell and E.R. Metcalf. 1983. Olfactory receptors in the melon fly

Dacus cucurbitae and the oriental fruit fly *Dacus dorsalis*. Proc. Nat. Acad. Sci. (USA) 80: 3143–3147.

Mitchell, W.A., R.L. Metcalf, E.R. Metcalf and S. Mitchell. 1985. Candidate substitutes for methyl eugenol as attractants for the area wide monitoring and control of the oriental fruit fly, *Dacus dorsalis* Hendel. Environ. Entomol. 14: 176–181.

Nakagawa, S., R.J. Prokopy, T.T.Y. Wong, J.R. Ziegler, S.M. Mitchell, T. Urago and E.J. Harris. 1978. Visual orientation of *Ceratitis captitata* flies to fruit models. Entomol. Exp. Appl. 24: 193–198.

Nakamura, K. and F. Kawasaki. 1977. The active space of *Spodoptera litura* (F.) sex pheromone and the pheromone component determining this space. Appl. Entomol. Zool. 11: 312–319.

Prokopy, R.J. and A.P. Economopoulos. 1971. Color responses of *Ceratitis capitata* flies. Zeit. Ang. Entomol. 80: 434–437.

Ripley, L.B. and G.A. Hepburn. 1935. Olfactory attractants for fruit flies. Entomol. Mem. Dept. Agr. S. Africa. 9: 3–17.

Smith, L.B., E.A. Richmond and P.A. van der Meulen. 1926. Geraniol as an attractant for insects, particularly the Japanese beetle. U.S. Pat. 1,527,568.

Sower, L.L., L.K. Gaston and H.H. Shorey. 1971. Sex pheromones of noctuid moths. XXVI. Female release rate, male response, threshold and communication distance for *Trichoplusia ni*. Ann. Entomol. Soc. Amer. 64: 1448–1456.

Sower, L.L., R.S. Kaae and H.H. Shorey. 1973. Sex pheromones of Lepidoptera. XLI. Factors limiting potential distance of sex pheromone communication in *Trichoplusia ni*. Ann. Entoml. Soc. Amer. 66: 1121–1122.

Stanley, B.H., H.E. Hummel and W.G. Ruesink. 1985. Estimating maximum horizontal areas of pheromone plumes. J. Chem. Ecol. 11: 1129–1140.

Steiner, L.F. 1952. Methyl eugenol as an attractant for the oriental fruit fly. J. Econ. Entomol. 45: 241–248.

Steiner, L.F. 1957. Low cost plastic fruit fly trap. J. Econ. Entomol. 50: 508–509.

Sutton, O.G. 1953. "Micrometeorology", McGraw-Hill, N.Y.

Visser, J.H. 1979. Electroantennogram responses of the Colorado beetle, *Leptinotarsa decemlineata*, to plant volatiles. Entomol. Exp. Appl. 25: 86–97.

Williams, M. 1986. The receptor from concept to function. Chapt. 21 in R.W. Egan ed. Annu. Rept. Medicinal Chem., Academic Press, N.Y.

Wilson, E.O., W.H. Bossert and F.E. Regnier. 1969. A general method for estimating threshold concentrations of odorant molecules. J. Insect. Physiol. 15: 597–610.

Wolf, W.W., A.N. Kishaba and H.H. Toba. 1971. Proposed method for determining density of traps required to reduce an insect population. Jour. Econ. Entomol. 64: 872–877.

Wright, R.H. 1958. The olfactory guidance of flying insects. Can. Entomol. 90: 81–89.

3

JAPANESE BEETLE

I. INTRODUCTION

The Japanese beetle *Popillia japonica* (Newman) (Coleoptera: Scarabaeidae) is native to Japan and was introduced into the United States on nursery stock shipped to New Jersey, about 1916. It has spread through almost all of the country east of the Mississippi River, into 26 states, and has been introduced several times into California. Its voracious feeding habits and very wide host range make it a very important insect pest of tree fruits, nursery stock, turfgrass, and home gardens.

II. LIFE HISTORY, APPEARANCE, AND HABITS

The Japanese beetle winters as a plump, white grub from 12–18 mm long in an earthen cell. The larva completes its growth during June and the adult emerges in late June and continues active feeding into September in the latitude of New Jersey. The adult Japanese beetle is metallic green to greenish bronze, 8–12 mm long, with reddish wing covers and two prominent and several smaller white spots near the tip of the abdomen. The adult beetles live an average of 39–52 days. The female lays an average of 40–60 white eggs in groups, 2 to 6 inches deep in the soil. These hatch in about two weeks, and egg laying begins in July and extends through September. The developing white grubs feed at first on decaying vegetation, but, when larger, on the fine roots of grasses and other plants. The larva passes through three instars over an average of about 136 days, attaining a length of about 20 mm. There is normally one generation annually, but in cold, wet seasons, the grubs may require two years to complete development (Fleming 1972).

A. Plants Attacked

Adult Japanese beetles are recorded as attacking 350 species of plants from more than 24 families, eating foliage, flowers, and fruits of tree fruits, berries, garden crops, and ornamental shrubs and trees. They also attack field crops such as alfalfa, clover, corn, and soybean. The plants most heavily attacked are shown in Table 3.1 (Fleming 1972). The adult beetles are particularly attracted to already infested foliage, and fruit and beetle aggregations or "balling" often occur on early ripening fruits of apple, apricot, cherry, peach, and grape. For example, Fleming (1972) recorded 296 beetles feeding on a single apple. A more scientific compilation of plant hosts based on controlled laboratory study of Japanese beetle feeding, where the relative palatability was determined by the weight of fecal pellets produced, is that of Ladd (1987). Those plants most heavily fed upon are shown in Table 3.2. Among 45 plant species from 24 families investigated, there was a 13-fold range in feeding, with *Vitis vinifera*, *Rosa* spp. and *Rubus idaeus* inducing feeding significantly greater than *Sassafras*, the standard. However, the foliage of 19 other species from Fleming's (1972) "heavily fed upon" category, induced feeding comparable to the *Sassafras* standard.

It is evident that the attraction of the Japanese beetle adults to volatile host plant odors is a major factor in host plant selection (Fleming 1972, Klein 1981). Thus the plant food preference of the beetles as shown in Tables 3.1 and 3.2 results from the common presence of specific plant kairomones. Volatiles isolated from a variety of attractive plants include (Fleming 1972, Sethi et al. 1976, Williams et al. 1982):

Volatile	*Source*
acetic acid	apple, peach, rose, sassafras
benzaldehyde	apple, peach
caproic acid	apple
citral	apple, rose
citronellol	rose
eugenol	rose, sassafras
geraniol	apple, grape, rose, sassafras
linalool	peach, rose, sassafras
2-phenylethanol	grape, rose
valeric acid	apple, peach, sassafras

III. IDENTIFICATION OF CHEMICAL ATTRACTANTS

The rapidity with which the newly introduced Japanese beetle spread throughout the eastern United States stimulated research work on chemical attractants, and within 3 years it was noticed that a variety of essential

Table 3.1. Plant Families and Species Most Severely Attacked by the Japanese Beetle*

Family	Species	Common name
Aceraceae	*Acer palmatium*	Japanese maple
	A. platanoides	Norway maple
Anacardiaceae	*Rhus toxicodendron*	poison ivy
Betulaceae	*Betula populifolia*	gray birch
Clethraceae	*Clethra alnifolia*	sweet pepper bush
Fagaceae	*Castanea dentata*	American chestnut
Graminaceae	*Zea mays*	corn
Hippocastaneaceae	*Aesculus hippocastaneum*	horse chestnut
Juglandaceae	*Juglans nigra*	black walnut
Lauraceae	*Sassafras albidum*	sassafras
Leguminaceae	*Glycine max*	soybean
Lillaceae	*Asparagas officinale*	asparagus
Lythraceae	*Lagerstroemia indica*	crepe myrtle
Malvaceae	*Althaea officinalis*	hollyhock
	A. rosae	hollyhock
	Hibiscus palustris	rose mallow
	H. syriacus	altheae
	Malva rotundifolia	mallow
Onagraceae	*Oenothera biennis*	evening primrose
Platanaceae	*Platanus acerifolia*	London planetree
Polygonaceae	*Polygonum orientalis*	smartweed
	P. pennsylvanicum	smartweed
	Rheum rhaponticum	garden rhubarb
Rosaceae	*Malus baccata*	crab apple
	M. sylvestris	apple
	Prunus americana	apricot
	P. avium	sweet cherry
	P. cerasus	sour cherry
	P. domesticus	plum
	P. persica	peach
	P. serotina	black cherry
	Rosa spp.	rose
	Sorbus americana	mountain ash
Salicaceae	*Populus nigra*	Lombardy poplar
	Salix discolor	pussy willow
Tillaceae	*Tilia americana*	linden
Ulmaceae	*Ulmus americana*	American elm
	U. procera	European elm
Vitaceae	*Parthenocissus quinquefolia*	Virginia creeper
	Vitis aestivalis	summer grape
	V. labrusca	fox grape
	V. vinifera	European wine grape

* Data from Fleming (1972).

Table 3.2. Plant Species Most Severely Attacked by Japanese Beetle*

Plant species	Common name	Family	Feeding index[1]
Rosa spp.	rose	Rosaceae	230
Vitus vinifera	grape	Vitaceae	165
Rubus idaeus	red raspberry	Rosaceae	135
Ulmus americana	European elm	Ulmaceae	127
Zea mays (silks)	sweet corn	Graminaceae	125
Acer platanoides	Norway maple	Aceraceae	124
Hibiscus syriacus	rose of Sharon	Malvaceae	124
Polygonum pennsylvanicum	smartweed	Polygonaceae	121
Brassica oleraceae	broccoli	Cruciferae	119
Acer palmatum	Japanese maple	Aceraceae	116
Tilia cordata	European linden	Tilaceae	115
Betula populiferum	gray birch	Betulaceae	114
Vitus labrusca	fox grape	Vitaceae	107
Malus baccata	crab apple	Rosaceae	107
Tilia americana	American linden	Tiliaceae	107
Parthenocissus quinquefolia	Virginia creeper	Vitaceae	106
Sorbus americana	mountain ash	Rosaceae	104
Quercus palustris	pin oak	Fagaceae	104
Sassafras albidum	sassafras	Lauraceae	100

* Data from Ladd (1987)

[1] weight of feces on plant/weight of feces on sassafras

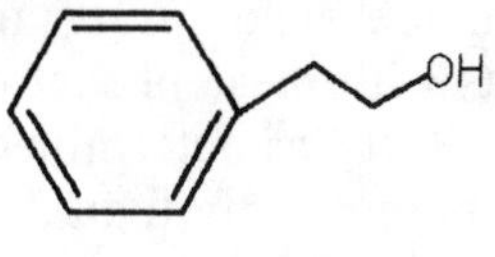

2-phenylethanol

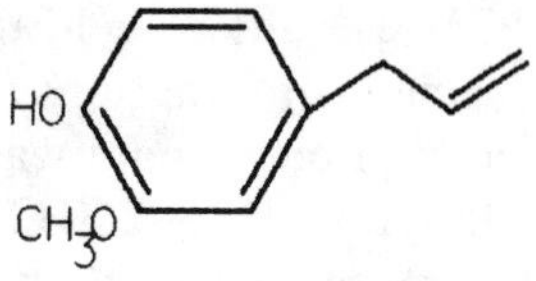

eugenol

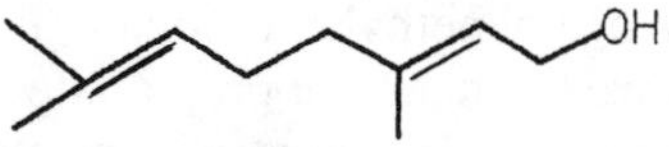

geraniol

oils and fruity odors from plants were effective attractants. Geraniol was found exceptionally attractive (Smith et al. 1926). Field trapping experiments by the United States Department of Agriculture (Richmond 1927) showed the relative attractivity of the more effective as:

Kairomone	Number of beetles trapped	Relative attractivity	
		Original	*Revised*
geraniol	10,071	100	100
eugenol	1,562	16	58
citronellal	1,214	12	—
citral	1,034	10	31
citronellol	620	6	—

However, the concentrations of these chemicals in the traps were varied, and revised values for relative attractivity (Fleming 1969) were determined as shown. These studies led to the use of mixtures of geraniol and eugenol generally standardized at a ratio of 10:1. This mixture was used as the standard attractant for field surveys of Japanese beetle populations by the U.S.D.A. until about 1940 (Fleming et al. 1940, Schwartz et al. 1970).

Subsequently, a mixture of anethole/eugenol (9:1) was found to be exceptionally attractive and was used from 1945–65 (Schwartz et al. 1970). Further evaluations showed a mixture of 2-phenethyl butyrate/eugenol (9:1) was more than twice as effective as the anethole/eugenol mixture, and the former was used in surveys after 1966 (Schwartz et al. 1970). However, 2-phenethyl acetate and 2-phenethyl propionate are more volatile than 2-phenethyl butyrate, and their higher release rates from field traps suggested the use of 2-phenethyl propionate/eugenol (7:3) (Ladd et al. 1973). Results of field trapping experiments with these phenethyl esters/eugenol combinations are shown in Table 3.3. It was determined that 2-phenethyl propionate was released 2 times more rapidly than 2-phenethyl butyrate at 25–26° C (McGovern et al. 1970) and that eugenol was slightly more volatile than 2-phenethyl butyrate.

Comparisons of the field attractivity of mixtures of phenethyl propionate/eugenol (Ladd et al. 1976) showed the following values for relative attractivity: (1:9) 115, (3:7) 136, (5:5) 134, (7:3) 100 (standard), and (9:1) 79. The relative attractivity of phenethyl propionate alone was 14, and of eugenol was 59. The mixture of phenethyl propionate/eugenol (3:7) was only 0.75 times as volatile as the (7:3) mixture and was subsequently adopted as the standard attractant mixture for U.S.D.A. and State survey programs (McGovern et al. 1973).

Caproic acid is an effective component of Japanese beetle volatile lures (Langford & Corey 1946) and mixtures of caproic acid/eugenol/phenethyl butyrate ((8:1:1), relative attractivity 284); and caproic acid/eugenol/phenethyl isovalerate ((8:1:1) attractivity 264), were more attractive in field trapping than geraniol/eugenol) ((9:1) standard, attractivity 100).

A three component lure of phenethyl propionate/eugenol/geraniol

Table 3.3. Field Catches of Japanese Beetles with 2-Phenethyl Esters and Eugenol (7:3)*

Lure	Average beetle catch	Relative attractivity
phenethyl acetate	308a	108
phenethyl propionate	285a	100
phenethyl butyrate	119b	42
phenethyl isovalerate	80c	28
phenethyl 2-methylvalerate	78c	27
phenethyl isobutyrate	67c	24
phenethyl butyrate/ eugenol (9:1)	57c	20
phenethyl 4-methylvalerate	46c	16
phenethyl 2-methylbutyrate	43c	15
phenethyl 2-ethylbutyrate	42c	15
phenethyl valerate	41c	14
phenethyl formate	40c	14
phenethyl hexanoate	39c	14
phenethyl pivalate	34c	11
control	13d	5

* Data from Ladd et al. (1973).

(3.5:3.5:3) was described by Ladd et al. (1975) as 1.5 times more attractive than the standard phenethyl propionate/eugenol (7:3) lure, and this is presently the lure most widely used.

There is a paucity of information about the relative attractivity of the individual kairomone volatiles evaluated at equivalent dosages under standard conditions. Fleming (1969) summarized the relative attractivity of a number of individual attractants in field tests compared with the standard geraniol/eugenol (10:1) mixture: (*S*)-β-citronellol 109, eugenol 70, isocaproic acid 35, valeric acid 29, *n*-caproic acid 25, 2-phenethanol 23, citronellol 16, citronellal 14, methyl salicylate 8, isoeugenol 6, 2-phenethyl acetate 5, safrole 5, anethol 4, and vanillin 2. Mixtures of these compounds were almost always more attractive to Japanese beetles than the expected sum of the attractivity of the individual compounds (Fleming 1969) and this olfactory synergism is illustrated in Table 3.4.

The positive attractant response of the Japanese beetle to individual plant volatiles such as caproic acid, geraniol, eugenol, citronellol, and 2-phenylethanol (phenethanol) and its esters that are widely distributed among the variety of host plants, demonstrates the presence of functional chemoreceptors specifically attuned to each of these kairomones. The range of attractant synergism, from 1.3–4.8 times shown by two, three, four, and even five component mixtures of these simple kairomones (Table 3.5) suggests that these receptors act in concert to produce maximal

Table 3.4. Attraction of Japanese Beetle to Mixtures of Volatile Plant Kairomones*

Mixture	Relative attractivity expected	Relative attractivity determined	Synergistic ratio
geraniol/caproic acid (1:4)	38	103	2.7
geraniol/eugenol (9:1)	79	100	1.3
geraniol/phenethanol (10:1)	74	98	1.3
citronellol/eugenol (10:1)	21	90	4.3
geraniol/phenethyl acetate (1:4)	20	84	4.2
geranyl acetate/eugenol (10:1)	39	75	1.9
phenethanol/eugenol (9:1)	28	74	2.6
citral/eugenol (10:1)	25	71	2.8
caproic acid/eugenol (9:1)	31	61	2.0
caproic acid/anethol (1:1)	22	58	2.7
isoamyl valerate/eugenol (9:1)	11	53	4.8
geraniol/anethol (1:9)	22	51	2.3
citronellol/eugenol (10:1)	19	45	2.4
geraniol/safrole (10:1)	73	40	0.55
geraniol/citral (1:9)	26	37	1.4

* Data from Fleming (1969).

Table 3.5. Mixtures of Plant Kairomones More than Twice as Attractive to the Japanese Beetle as Geraniol/Eugenol (10:1)*

Mixture	Relative attractivity
geraniol/eugenol/phenethyl isovalerate (1:1:8)	370
caproic acid/eugenol/phenethyl butyrate (18:1:1)	301
anethol/caproic acid/eugenol/phenethyl butyrate/isovaleric acid (8:8:3:3:8)	285
anethol/caproic acid/eugenol/phenethyl butyrate (9:9:1:1)	279
caproic acid/eugenol/phenethyl isovalerate (8:1:1)	264
anethol/caproic acid/eugenol (9:9:2)	246
caproic acid/eugenol/geraniol/phenethyl butyrate (4:1:4:1)	232
geraniol/eugenol/phenethyl butyrate (1:1:8)	222
anethol/caproic acid/eugenol/geraniol/phenethyl butyrate (6:5:1:1:1)	206
geraniol/eugenol (10:1)	100
geraniol/eugenol (5:1)	142
geraniol/eugenol (5:2)	146
geraniol/eugenol (5:4)	165
geraniol/eugenol (1:1)	153

* Data from Fleming (1969).

behavioristic responses to mixtures of volatiles that approximate those elaborated by favored host plants. The retention of functional sensory receptors and of the insect's ability to integrate their collective depolarization after millions of years of plant/insect evolution is suggested as the basis for the remarkable facility with which the Japanese beetle has developed its very wide host range.

Although much of the successful development of kairomone lures for the Japanese beetle has resulted from serendipity, a great deal remains to be learned from careful analysis of host plant volatiles and the behavioristic responses of the Japanese beetle to these, both singly and in combination.

IV. PARAKAIROMONES

Eugenol, geraniol, and phenethanol are natural products plant kairomones for the Japanese beetle adult. In continuing efforts to find improved lures for this insect, structure-activity studies have been conducted by the U.S.D.A. with chemicals having structural similarities to both phenethanol esters and to eugenol.

A. Phenethanol Analogues

Substituted cyclohexane carboxylic acid esters alone were unattractive. However, a mixture of methylcyclohexane propionate/eugenol (1:9) was 2.5 times more attractive than phenethyl butyrate/eugenol (1:9) (Table 3.6) (McGovern et al. 1970). The unsaturated propenoic acid ester of phenethanol (phenethyl propenoate) in 7:3 mixture with eugenol had a relative attractivity of 121 compared to 100 for the standard phenethyl propionate/eugenol (7:3) (McGovern & Ladd 1981). Cycloalkanoate esters of phenethanol were investigated and phenethyl cyclopropanoate/eugenol (1:9 and 3:7) was slightly more attractive to Japanese beetle adults than the standard. The cyclohexanoic, cyclobutanoic, and cyclopentanoic esters were of low attractivity (McGovern & Ladd 1981). Replacement of the phenyl moiety with cyclohexyl gave attractive mixtures with eugenol, and 2-cyclohexylethyl propionate/eugenol (7:3) had a relative attractivity of 97 compared to 100 for the standard phenethyl propionate/eugenol (7:3) mixture. Varying the ratios of the components produced only minor changes in attractivity (McGovern & Ladd 1981). The interesting demonstration that replacement of the planar phenyl ring by the chair-like staggered cyclohexyl ring did not result in a pronounced loss of attractivity in admixture with eugenol, was explainable in terms of decreased attraction of the cyclohexyl esters, but a greater volatility and increased release rate (McGovern & Ladd 1981).

Table 3.6. Attraction of Japanese Beetle to Mixtures of Cyclohexanecarboxylic Acid Esters with Eugenol (9:1)

Mixture	Average trap catch	Relative attractivity
phenethyl butyrate/eugenol		100
methyl cyclohexanecarboxylate/eugenol	151	18
methyl cyclohexaneacetate/eugenol	777	91
methyl cyclohexanepropionate/eugenol	3557	255
methyl cyclohexanebutyrate/eugenol	1854	133
ethyl cyclohexanebutyrate/eugenol	369	26
propyl cyclohexanebutyrate/eugenol	129	93
ethyl cyclohexanepropionate/eugenol	2074	148
propyl cyclohexanepropionate/eugenol	447	32
butyl cyclohexanepropionate/eugenol	238	17
methyl 3-cyclohexenecarboxylate/eugenol	178	30
methyl 2-methylcyclohexanecarboxylate/eugenol	128	21
ethyl 2-methylcyclohexanecarboxylate/eugenol	132	10
propyl 2-methylcyclohexanecarboxylate/eugenol	243	19
methyl 6-methyl-3-cyclohexenecarboxylate/eugenol	193	32
ethyl 6-methyl-3-cyclohexenecarboxylate/eugenol	229	166
propyl 6-methyl-3-cyclohexenecarboxylate/eugenol	47	34

* Data from Mcgovern et al. (1970)

B. Eugenol Analogues

The eugenol component of Japanese beetle attractants is very important, as eugenol is more attractive than phenethyl propionate (Ladd et al. 1976). Investigations of phenylpropanoids related to eugenol in admixture with phenethyl propionate demonstrated the critical nature of the 3-methoxy-4-hydroxyphenyl substitution characteristic of eugenol (McGovern & Ladd 1984). The analogue of eugenol with a saturated side chain (2-methoxy-4-propylphenol (3-methoxy-4-hydroxy-1-propylbenzene)) was approximately as attractive (relative attractivity 79) as eugenol, (relative attractivity 100), when used in admixture with phenethyl propionate (7:3). When the ratios were reversed, phenethyl propionate/2-methoxy-4-propylphenol (3:7) had a relative attractivity of 119, compared to 100 for phenethyl propionate/eugenol (3:7). Other eugenol analogues (isoeugenol, methyl eugenol, estragole, and anethole) used with phenethyl propionate had relative attractivity values ranging from 4–10, indicating the importance of the 3-methoxy-4-hydroxyphenyl configuration for olfactory response by the Japanese beetle (McGovern & Ladd 1984). Used alone, the 2-methoxy-4-propylphenol (relative attractivity 52) was slightly more attractive than eugenol (relative attractivity 28). This is probably the result of its slightly higher volatility (McGovern & Ladd 1984).

These investigations of parakairomones have developed new synthetic lures highly attractive to both sexes of the Japanese beetle. The most effective new mixtures that are available as potential tools for insect management programs are: phenethyl propenoate/eugenol (7:3), phenethyl cyclopropanoate/eugenol (1:9), 2-cyclohexylethyl propionate/eugenol (7:3)(McGovern & Ladd 1981), and phenethyl propionate/2-methoxy-4-propylphenol (3:7)(McGovern & Ladd 1984).

V. KAIROMONES FOR MONITORING AND CONTROLLING THE JAPANESE BEETLE

Hundreds of thousands of Japanese beetle traps are baited with kairomone lures each summer in the eastern United States, and billions of beetles are trapped and killed. These traps are effective in monitoring for the presence of Japanese beetles, to destroy the females before they lay eggs in the soil, and to protect valuable ornamental, nursery, orchard, and turf crops. However, Japanese beetle populations are often so great that even 100 traps per mi^2 will cause only minimal reductions in the overall beetle population (Fleming et al. 1940).

The basic trap has been little changed over the past 45 years. It consists of 4 plastic baffles arranged about 4 in above a funnel which feeds the arriving beetles directly into a constricted plastic bag or glass jar (Figure 2.1). Bright yellow, glossy baffles are more attractive than green, blue, red, white or aluminum colors (Fleming et al. 1940). The attractant most generally used is a mixture of eugenol/phenethyl propionate/geraniol impregnated into a plastic strip for slow release. There are several commercial traps available to the householder, and a popular variety contains 1 g each of phenethyl propionate and geraniol and 2.4 g of eugenol, which is impregnated in a plastic strip for slow release. This lure is stated to remain effective for the 4 month summer season when Japanese beetles are present. Such lure baited traps are stated to attract about equal numbers of male and female beetles from distances of about 300 to 500 yards (Fleming et al. 1940).

Klein et al. (1981) have shown that incorporation of the female sex pheromone of the Japanese beetle, (*R*,*S*)-5-(1-decenyl)-dihydro-2-(*3H*)-furanone together with the kairomone lures significantly improves the catch of male Japanese beetles as shown in Table 3.7. In the commercial trap described above, a small plastic strip containing 1.0 mg of the female pheromone is supplied.

According to Fleming et al. (1940) these kairomone baited traps are most effective when hung at a height of about 4 to 5 feet above the ground, and 10 to 25 feet to the windward side of the property to be protected. In orchards, the traps should be placed 10 to 25 feet to the prevailing

Table 3.7. Attraction of Japanese Beetles to Kairomone Lures Fortified with Female Sex Pheromone*

Lure	Average beetle catch per trap		
	Male	Female	Total
phenethyl propionate/eugenol (3:7)	304c	340c	652c
pheromone[1]	1122b	64b	1186b
phenethyl propionate/eugenol (3:7) plus pheromone	1845a	662a	2507a

* Data from Klein et al. (1981)

[1] (*R,S*)-5-(1-decenyl)-dihydro-2-(*3H*)-furanone

windward direction from the trees to be protected. A properly baited trap should capture at least 70% of all the beetles in the area. Japanese beetles are most active and responsive to lures from 9 a.m. to 6 p.m., and at temperatures of 80 to 90° F. At night, and during cool and rainy weather, the beetles remain on the ground or on plants and cannot be captured. A single trap is adequate for a suburban lot, but traps should be placed 100 to 200 ft apart on large properties. Trapping should be begun when the first beetles emerge in the early summer and continued as long as the beetles are present (Fleming et al. 1940).

The use of kairomone lures for trapping Japanese beetles has had belated recognition as an important method of suppression of this pest, and was not recommended by the U.S.D.A. until Schwartz (1975) endorsed trapping as a non-insecticidal method for destroying beetles and reducing damage. The state of Maryland (Corey & Langford 1955) recommended trapping and reported that more than 369 tons of adult beetles were destroyed during a single summer. Hamilton et al. (1971) reported that mass trapping in Nantucket, Massachusetts reduced Japanese beetle populations by at least 50% over a period of three years. The employment of 1400 traps at Dulles International Airport was reported to have eliminated the need to treat departing aircraft to eliminate "hitchhiking" Japanese beetles (Klein 1981). Trapping in areas of isolated infestations has been effective in preventing establishment of the beetles for several years.

REFERENCES

Cory, E.N. and G.S. Langford. 1955. The Japanese beetle retardation program in Maryland. Univ. MD Ext. Bull. 156, 20 pp.

Fleming, W.E. 1969. Attractants for the Japanese beetle. U.S.D.A. Tech. Bull. 1399, 87 pp.

Fleming, W.E. 1972. Biology of the Japanese Beetle. U.S.D.A. Tech. Bull. 1449.

Fleming, W.E., E.D. Burgess, and W.W. Maine. 1940. The use of traps against the Japanese beetle. U.S.D.A. Cir. 594, 12 pp.

Hamilton, D.W., P.H. Schwartz, B.G. Townshend and C.W. Jester. 1971. Traps reduce an isolated infestation of Japanese beetle. J. Econ. Entomol. 64: 150

Klein, M.G. 1981. Mass trapping for suppression of Japanese beetles, pp. 183–190, in E.R. Mitchell, ed. "Management of Insect Pests with Semiochemicals.", Plenum, NY.

Klein, M.G., J.H. Tumlinson, T.L. Ladd, Jr. and R.E. Doolittle. 1981. Japanese beetle (Coleoptera: Scarabaeidae): response to synthetic sex attractant plus phenethyl propionate: eugenol. J. Chem. Ecol. 7: 1–7

Ladd, T.L. Jr. 1987. Japanese beetle: influence of favored food plants on feeding response. J. Econ. Entomol. 80: 1014–1017.

Ladd, T.L. Jr, and T.P. McGovern. 1980. Japanese beetle: a superior attractant phenethyl propionate + eugenol + geraniol 3:7:3. Jour. Econ. Entomol. 73: 689–691.

Ladd, T.L. Jr., T.P. McGovern, M. Beroza, B.G. Townshend, M.C. Klein and K.O. Lawrence. 1973. J. Econ. Entomol. 66: 369–370.

Ladd, T.L. Jr., C.R. Bariff, M. Beroza, and T.P. McGovern. 1975. Japanese beetles: attractancy of mixtures of lure containing phenethylpropionate and eugenol. J. Econ. Entomol. 68: 819–820.

Ladd, T.L. Jr., T.P. McGovern, M. Beroza, C.R. Buriff, and M.G. Klein. 1976. Japanese beetles: attractancy of phenethylpropionate + eugenol (3:7) and synthetic eugenol. J. Econ. Entomol. 69: 468–470.

Langford, G.S. and E.N. Cory. 1946. Japanese beetle attractants with special reference to caproic acid and phenyl ethyl butyrate. J. Econ. Entomol. 39: 245–247.

McGovern, T.P., M. Beroza, P.H. Schwartz, D.W. Hamilton, J.C. Ingangi, and T.L. Ladd. 1970. Methyl cyclohexanepropionate and related chemicals as attractants for Japanese beetles. J. Econ. Entomol. 63: 276–280.

McGovern, T.P., M. Beroza, and T.L. Ladd. 1973. Phenethyl propionate and eugenol, a potent attractant for the Japanese beetle (*Popillia japonica* Newman). U.S. Pat. 3,761,584, Sept. 25.

McGovern, T.P. & T.L. Ladd Jr. 1981. Japanese beetle: new synthetic attractants. J. Econ. Entomol. 74: 194–196.

McGovern, T.P. and T.L. Ladd, Jr. 1984. Japanese beetle attractant: tests with eugenol substitutes and phenethyl propionate. J. Econ. Entomol. 77: 370–373.

Richmond, E.A. 1927. Olfactory response of the Japanese beetle (*Popillia japonica* Newm.). Proc. Ent. Soc. Washington 29: 36–44.

Schwartz, P.H. jr. 1975. Control of insects on deciduous fruits and tree nuts in the home orchard without insecticides. U.S.D.A. Home & Garden Bull. 211, 36 pp.

Schwartz, P.H., D.W. Hamilton, C.W. Jester, and B.G. Townshend. 1966. Attractants for Japanese beetles tested in the field. J. Econ. Entomol. 59: 1516–1517.

Schwartz, P.H., L.A. Hickey, D.W. Hamilton, and T.L. Ladd, Jr. 1969. Combinations of insecticides and baits for the Japanese beetle. J. Econ. Entomol. 62: 738–740.

Schwartz, P.H. Jr., D.W. Hamilton, and B.G. Townshend. 1970. Mixtures of compounds as lures for the Japanese beetle. J. Econ. Entomol. 63: 41–43.

Sethi, M.L., G.S. Rao, B.K. Choudhury, J.F. Morton, and G.J. Kapadia. 1976. Identification of volatile constituents of *Sassafras albidum* root oil. Phytochemistry 15: 1773–1775.

Smith, L.B., E.A. Richmond, and P.A. Vander Meulen. 1926. Geraniol as an attractant for insects, particularly the Japanese beetle. U.S. Pat. 1,572,568.

Williams, P.J., C.R. Strauss, B. Wilson, and R.A. Massey-Westropp. 1982. Use of C_{18} reversed phase liquid chromatography for the isolation of monoterpene glycosides and non-isoprenoid precursors from grape juice and wines. J. Chromat. 235: 471–481.

4

DIABROTICITE ROOTWORM BEETLES

I. INTRODUCTION

The Chrysomelidae, or leaf beetles, comprise about 20,000 described species of Coleoptera that feed almost exclusively on plants. The tribe Luperini (subfamily Galerucinae) exhibits a remarkable coevolutionary association with the plant family Cucurbitaceae and this provides the most comprehensive example of the role of plant allelochemicals acting as kairomones to promote host selection and feeding by phytophagous insects. The Luperini is composed of about 480 species of Old World Aulacophorites and 900 species of New World Diabroticites (Krysan & Smith 1987, Smith & Lawrence 1967, Wilcox 1972). Maulik (1936) cites the remarkable similarities between major genera of the two groups: "In the old world, *Aulacophora* represents *Diabrotica* . . . In larval, pupal, and adult structures, in breeding habits, and in food plants there is a remarkable resemblance between the two genera."

The Aulacophorites and the Diabroticites include some of the world's most destructive insects as shown in Table 4.1. As the common names of the most injurious species suggest, e.g. plain pumpkin beetle *Aulacophora abdominalis*, red pumpkin beetle *A. foveicollis*, pumpkin beetle *A. hilaris*, cucurbit leaf beetle *A. femoralis*, banded cucumber beetle *Diabrotica balteata*, spotted cucumber beetle *D. u. howardi*, cucurbit beetle *D. speciosa*, saddled cucumber beetle *D.* (*Paranapiacaba*) *connexa*, painted cucumber beetle *D. picticornis*, striped cucumber beetles *A. trivittatum* and *A. vittatum*: these Luperini species are closely associated with host plants of the Cucurbitaceae (Metcalf, 1985). Host plant records for the Luperini are regrettably sparse but at least 65 species have been collected from the blossoms and foliage of Cucurbitaceae (Table 4.1), representing about 90% of the published host plant records (Metcalf 1985). The magnitude of this association could be greatly expanded by further investigation, as 12 species of Aulacophorina were collected from wild and cultivated Cucurbitaceae in Taiwan (Takizawa 1978) and 16 species

Table 4.1. Luperini Beetles (Chrysomelidae: Galerucinae) Associated with Cucurbitaceae*

Species	Plant hosts	Locations
Aulacophorina		
Aulacophora spp.		
abdominalis (F.)	cucumber, melon, pumpkin	Indonesia, Australia
atripennis (F.)	gourd, muskmelon, pumpkin	Indonesia
bicolor (Weber)	Cucurbitaceae	Taiwan, Indonesia, China, India, Japan, Philippines
cincta (F.)	snake, bitter, bottle gourds	India, Sri Lanka
coffeae (Hornstedt)	melon, pumpkin	Indonesia, Fiji
excavata Baly	*Citrullus, Cucumis, Cucurbita, Luffa*	India
femoralis (Mots.)	Cucurbitaceae	Asia
foveicollis (Lucas)	pumpkin, squash, muskmelon	Asia, Africa, S. Europe
hilaris (Boisd.)	pumpkin, marrow	Australia, Micronesia
lewisii Baly	Cucurbitaceae	S.E. Asia, Pacific Islands
loochoonensis Chujo	Cucurbitaceae	Ryukyu Islands
nigripennis Mots.	Cucurbitaceae	Siberia, China, Japan, Korea, Taiwan
olivieri Baly	cucumber, melon, squash, melon	Australia
quadrimaculata (F.)	*Citrullus lanatus, Cucurbita pepo*	Australia, Pacific Islands
similis (Olivier)	*Citrullus lanatus, Cucumis melo, C. sativa, C. pepo*	S.E. Asia
Paridea spp.		
angulicollis (Mots.)	*Gymnostemma pentaphyllum, Trichosanthes cucumeroides*	China, Japan, Taiwan
costata (Chujo)	wild Cucurbitaceae	China, Taiwan
sauteri (Chujo)	wild Cucurbitaceae	Taiwan
sexmaculata (Laboissiere)	wild Cucurbitaceae	Taiwan
testacea Gressit & Kimoto	wild Cucurbitaceae	China, Taiwan
Agetocera spp.		
discedens Weise	wild Cucurbitaceae	Taiwan
taiwana Chujo	wild Cucurbitaceae	Taiwan
Lamprocopa spp.		
delata (Erichson)	Cucurbitaceae	Ethiopia, Angola, Zambesi
Paragetocera spp.		
involuta Laboissiere	wild Cucurbitaceae	China, Tibet, Taiwan

Table 4.1 (continued). Luperini Beetles (Chrysomelidae: Galerucinae) Associated with Cucurbitaceae*

Species	Plant hosts	Locations
Diabroticina		
Diabrotica spp.		
amecameca Krysan & Smith	Cucurbitaceae cultivars	Mexico, Central America
balteata LeConte	Cucurbitaceae	United States, Venezuela, Columbia
barberi Smith & Lawrence	*Cucurbita foetidissima*	Central and Eastern United States
carolae Krysan & Smith	*Cucurbita*	Peru
cristata (Harris)	*Cucurbita*	Central United States
graminea Baly	*Cucurbita moschata*	Central America, W. Indies
hilli Krysan & Smith	*Cucurbita ficifolia*	Mexico, Guatemala
lemniscata LeConte	*Cucurbita foetidissima*	United States, Mexico
linsleyi Krysan & Smith	*Cucumis melo*	Mexico
longicornis (Say)	*Cucurbita foetidissima*, cultivars	United States, Mexico
mapiriensis Krysan & Smith	Cucurbitaceae cultivars	Peru
porracea Harold	*Cucurbita pepo, C. foetidissima*	Mexico, Central America
sicuanica (Bechyne)	Cucurbitaceae	Peru
speciosa (Germar)	*Cucurbita andreana*	Mexico to Argentina
tibialis Jacoby	Cucurbitaceae	S.W. United States, Mexico
virgifera virgifera LeConte	*Cucurbita foetidissima*	Central United States
virgifera zea Krysan & Smith	*Cucurbita foetidissima, C. pepo*	S.W. United States, Mexico
viridula (F.)	*Cucurbita gracilor, C. moschata*	Mexico, Central America, S. America
undecimpunctata howardi Barber	Cucurbitaceae	United States
undecimpunctata undecimpunctata Mann.	Cucurbitaceae	W. United States

Table 4.1 (continued). Luperini Beetles (Chrysomelidae: Galerucinae) Associated with Cucurbitaceae*

Species	Plant hosts	Locations
Acalymma spp.		
annulatum (Suffrian)	Cucurbitaceae	Cuba, Puerto Rico
bivitattum (F.)	*Cucumis melo, C. sativa*	Cuba, Guiana, Brazil
bivittulum amazonum Bechyne	*Cucurbita maxima*	S. America
bivittulum bivittulum (Kirsch)	*Cucurbita maxima*	Argentina, Brazil, Bolivia, Peru
bivittulum exiguum Bechyne & Bechyne	*Cucurbita maxima*	Ecuador, Peru
blandulum (LeConte)	*Cucurbita foetidissima*	United States
coruscum (Harold)	*Cucurbita pepo, C. maxima*	Columbia, Venezuela
gouldi Barber	*Echinocystis lobata, Sicyos angulatus*	United States
innubum (F.)	*Cucumis, Cucurbita*	Puerto Rico, Dominica, Venezuela
isogenum Bechyne & Bechyne	*Cucurbita maxima*	Venezuela
peregrinum (Jacoby)	*Cucumis melo, Marah*	S. Texas, Mexico
punctatum punctatum Bechyne	*Cucurbita maxima*	Brazil
punctatum coremum Bechyne	*Cucurbita maxima*	Venezuela
separatum (Baly)	Cucurbitaceae	Guiana
thiemei (Baly)	Cucurbitaceae	Haiti, Jamaica
trivittatum (Mann.)	Cucurbitaceae	W. United States
vinctum (LeConte)	*Cucurbita okeechobeensis*	Florida, Georgia
vittatum (F.)	Cucurbitaceae	United States
Paranapiacaba spp.		
connexa (LeConte)	Cucurbitaceae	S.W. United States, Mexico
tricincta Say	Cucurbitaceae	S.W. United States

* Data from Krysan & Smith 1987, Krysan et al. (1984), Maulik (1936), Metcalf (1985), Metcalf & Lampman (1989a), Smith & Lawrence (1967), Takizawa (1978), and Wilcox (1972).

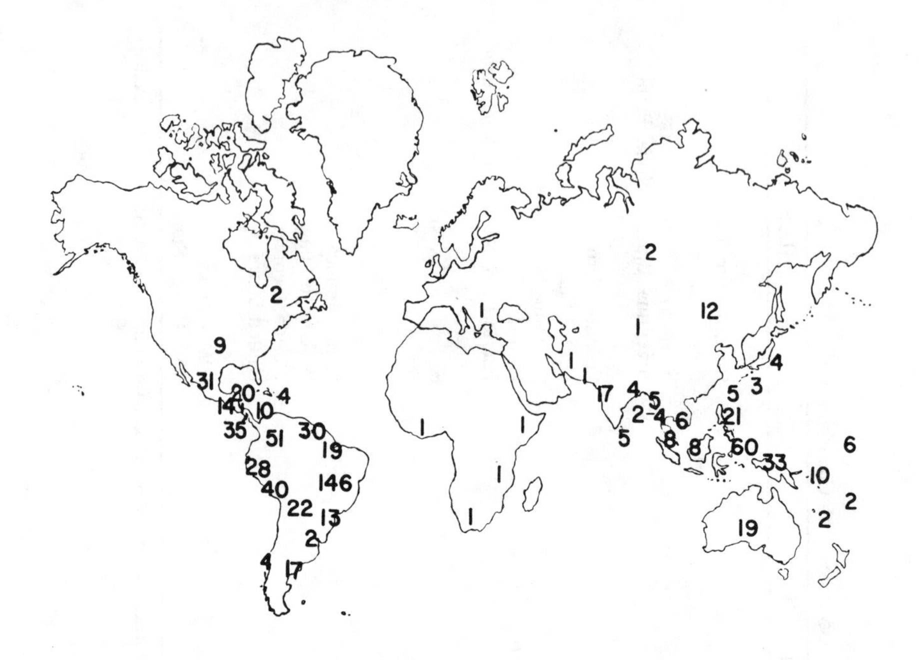

Figure 4.1. Distribution and numbers of described species of Diabroticina in the New World and Aulacophorina in the Old World. (Data from Wilcox 1972). Reprinted with permission from Metcalf (1985).

of *Diabrotica* were collected from the foliage and flowers of Cucurbitaceae in Peru (Krysan et al. 1984).

The genus *Diabrotica*, consisting of at least 338 valid species, is largely neotropical, and about 300 species are in the *fucata* group primarily confined to South and Central America (Figure 4.1) while the *virgifera* subgroup of about 35 species are found in North America (Krysan & Smith 1987). The species of the *fucata* group are polyphagous as adults and larvae, feed on plants of at least eight families, and are typically multivoltine and overwinter as adults in the southern United States. The *virgifera* species are usually univoltine and overwinter in the northern hemisphere as diapausing eggs. The majority of species in this subgroup feed as larvae on grasses and are polyphagous pollen feeders as adults (Krysan & Smith 1987). Several of the members of the *virgifera* group are extremely damaging pests of corn, *Zea mays*, e.g. the northern corn rootworm *D. barberi*, the western corn rootworm *D. v. virgifera*, and the cucurbit beetle *D. speciosa*. However, the proclivity of *Diabrotica* species to be rootworm feeders on corn does not affect the adult preferences for cucurbits, as most species of the *virgifera* group have been collected from the blossoms of Cucurbitaceae (Krysan & Smith 1987, Krysan et al. 1984).

The Chrysomelidae are believed to have originated in the upper Jurassic at about the time of the breakup of Gondwana into South America, Africa, Australia, and India (Crowson 1981). This provides a plausible explanation for the remarkable zoogeographic similarity in host-plant affinities of the Aulacophorites (Old World) and Diabroticites (New World). Speciation within the Diabroticites is thought to have occurred over the past 30 million years (0.7 genetic distance) since *Acalymma* separated from *Diabrotica* and the latter evolved into the polyphagous *fucata* group of non-diapausing Neotropical species and the stenophagous *virgifera* group of diapausing Nearctic species (Krysan et al. 1989).

II. LIFE HISTORIES, APPEARANCE, HABITS

The Luperini beetles are typically small (3–5 mm), ovate in outline, and often spotted or striped in strong colors. The larvae are yellowish with brown heads and tunnel into the roots of various plants. The important pest species of the *fucata* group in temperate North America are the banded cucumber beetle *Diabrotica balteata*, the two-spotted cucumber beetle *D. undecimpunctata howardi*, and the western spotted cucumber beetle *D. u. undecimpunctata*. These species overwinter as adults in the southern United States among crop remnants and become active early in the spring when the temperature reaches 70° F. They fly about, mate, and the females lay their 500–1000 eggs singly in the ground around the bases of cucurbits, sweet potato, corn, and soybeans. The whitish larvae tunnel into the roots, passing through four instars over a period of about 30 days. They pupate in the soil and the adults emerge during early July. The spotted cucumber beetle has two generations a year in the southern part of the range and a partial second generation in the northern areas (Metcalf et al. 1962).

The *virgifera* group of species (about 35) are usually univoltine and overwinter as diapausing eggs. The important crop pests are the northern corn rootworm *Diabrotica barberi* and the western corn rootworm *D. virgifera virgifera*. These two species are stenophagous, the larvae attacking the roots of grasses, especially corn. *D. barberi* adults are uniformly pale green to yellow-green and about 5 mm long. They emerge from pupae in the soil about the last week in July and feed on corn silk and pollen. The typical female lays about 300 eggs about the roots of corn, and these remain in diapause until spring when the newly emerged larvae migrate to the nearest corn roots where they tunnel and pass through four instars over a period of about one month. The larger larvae sever many of the fine rootlets of the plant and cause the plants to fall over or "lodge" after heavy rain and wind. There is a single generation each year, but about 5% of the overwintering eggs may undergo extended diapause through an additional year.

D. v. virgifera adults are slightly larger, about 6 mm long, and are yellow brown with three dark longitudinal bands on the elytra. The life history is very similar to that of the northern corn rootworm, the adult female emerging from soil pupation in late July and laying as many as 1000 eggs in the soil about corn plants. The larvae develop primarily on the roots of corn, and the adults feed heavily on corn silk and pollen.

The striped cucumber beetle, *Acalymma vittatum*, represents a third type of life history. The adult beetles are about 6 mm long, yellow-green in color, and have three prominent longitudinal stripes. Unmated adults hibernate in woodlands, hedges, and fence rows in direct contact with the soil. The adult beetles become active in April or early May when the temperature reaches 55–60° F and feed on pollen, petals, and leaves of willow, hawthorne, thorn apple, buckeye, apple, elm, syringa, and related plants. When young cucurbit plants emerge from the ground, the beetles settle on them, devouring cotyledons, leaves, and stems. The females lay yellow-orange eggs about the base of squash, cucumbers, and melons–often in cracks in the ground. The developing larvae tunnel through and often devour the roots over a period of 2–6 weeks, pupate in the soil, and emerge as adults during July and August. They feed heavily on cucurbit leaves and blossoms, and later in the fall on the fruits of cucurbits. There is only a single generation in northern areas, and two or more in the South. The striped cucumber beetle and its western relative, *A. trivittatum*, are efficient vectors of cucurbit wilt caused by *Erwinia tracheiphila* (Metcalf et al. 1962).

III. ROLE OF CUCURBITACINS IN HOST PLANT ASSOCIATIONS OF LUPERINI

The plant family Cucurbitaceae contains more than 900 species in about 100 genera, many familiar as the gourds, squash, cucumbers, and melons of *Cucurbita*, *Cucumis*, *Citrullus*, *Lagenaria*, *Marah*, *Sicyos*, *Echinocystis*, *Ecbalium*, and *Bryonia*. These plants are uniquely characterized by their biosynthesis of a group of more than 20 oxygenated tetracyclic triterpenoids, the cucurbitacins (Cucs) (Figure 4.2). These extremely bitter and highly toxic secondary plant compounds are responsible for the characteristic bitter taste of most wild Cucurbitaceae (Rehm 1960, Lavie & Glotter 1971, Metcalf 1986) (Section III-D).

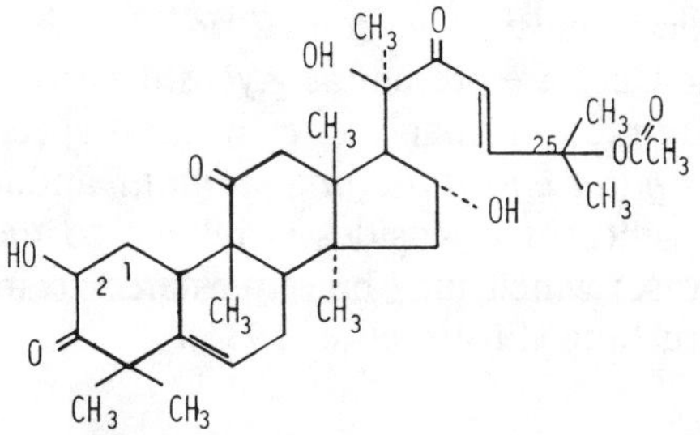

Figure 4.2. Structure of cucurbitacin B (Cuc B). Cuc D is C_{25}-OH; Cuc E is $C_1 = C_2$; Cuc I is $C_1 = C_2$, C_{25}-OH; Cuc F is C_2-OH, C_3-OH, C_{25}-OH; Cuc G is C_{24}-OH, C_{25}-OH; Cuc L is $C_1 = C_2$, C_{23}-C_{24}, C_{25}-OH.

A. Chemical Nature of Cucurbitacins

The cucurbitacins (Cucs) are triterpenoids biosynthesized in plants from mevalonic acid. While found principally in the Cucurbitaceae, their presence has also been recorded in a few genera of related plant families, e.g. Begoniaceae, Brassicaceae, Datisceae, all of the superorder Violofloreae and in Euphorbiaceae, Rosaceae, and Scrophulariaceae (Dryer & Trousdale 1978). For example, Cucs have been identified in 16 species of *Iberis* (Brassicaceae) (Lavie & Glotter 1971, Curtis & Meade 1971).

At least 20 chemically different cucurbitacins have been identified from plants (Lavie & Glotter 1971, Guha & Sen 1975) (Figure 4.2). In the Cucurbitaceae, Cuc B is the predominant form found in about 91% of all species characterized, followed by Cuc D (69%), Cucs G and H (47%), Cuc E (42%), Cuc I (22%), Cucs J and H (9%), and Cuc A (7%). Cucs C, F, and L have been found only in single species (2%) (Rehm et al 1957). The two primary Cucs are B and E, and the other Cucs are formed by enzymatic processes occurring during plant development and maturation (Lavie & Glotter 1971, Rehm 1960). Cucurbitacin B can be metabolized to Cucs A, C, D, F, G, and H and is characteristic of *Coccinia*, *Cucumis*, *Lagenaria*, and *Trichomeria* (Rehm et al. 1957). Cuc E can be metabolized in a similar way to Cucs I, J, K, and L and is characteristic of *Citrullus*. *Cucurbita* contains two discrete groups of species characterized by the presence of either Cuc B or Cuc E (Metcalf et al. 1982). Cuc B is converted to Cuc E by Cuc Δ^1 dehydrogenase, which forms the diosphenol grouping $C_1 = C_2$; Cuc Δ^{23} reductase converts the $C_{23} = C_{24}$ Cuc B and E series into the dihydrocucurbitacins. Cucs B and E are metabolized to Cucs D and I through deacetylation by Cuc acetylesterase (Schwartz et al. 1964, Schabort & Teijema 1968, Lavie & Glotter 1971).

In *Cucumis*, *Lagenaria*, and *Acanthosicyos* the Cucs are present as free aglycones. However, in most species of *Citrullus, Echinocystis, Coccinea,* and *Peponium,* the Cucs are present as glycosides (Rehm et al. 1957). In *Cucurbita,* Cucs are present as aglycones in most species, but glycosides are found in *C. cylindrata, foetidissima, palmata,* and *texana* (Metcalf et al. 1982). The presence of glycosides is related to the absence of a β-glucosidase (elaterase) which may be sequestered in intact plant tissues and released by crushing (Enslin et al. 1956).

B. Distribution of Cucurbitacins in Plants

Cucs from the Cucurbitaceae are found in all parts of the plant: roots, stems, leaves, fruits, and occasionally in the seeds.

Roots. The concentration in the roots increases with age and reaches high levels in perennials, e.g. 1.4% in *Citrullus naudineanus,* 1.1% in *Acanthosicyos horridus,* and 0.9% in *Colocynthis ecirrhosa* (Rehm 1960). In the roots of 18 species of *Cucurbita,* Cucs B-D were found in seven, up to 0.43% in *C. ecuadorensis,* and Cucs E-I in six, up to 0.38% in *C. palmata* (Metcalf et al. 1982).

Leaves. Young rapidly growing leaves of *Colocynthis* (*Citrullus*) *vulgaris* and *C. ecirrhosa* contained only about 0.01% Cucs, but the concentration increased to 0.1–0.3% by the end of the vegetative period (Rehm 1960). In 18 species of *Cucurbita,* Cucs B-D were found in the leaves of seven species, up to 0.059% in *C. lundelliana,* and Cucs E-I were found in six species, up to 0.1% in *C. okeechobeensis* (Metcalf et al. 1982).

Fruits. The fruits of wild Cucurbitaceae contain high concentrations of Cucs, 0.1% in *Citrullus colocynthis* and *C. eccirhosa* (Cuc E), *Cucumis angolensis* and *C. longipes* (Cuc D), *C. myriocarpus* (Cuc A), and *C. sativus* (Cuc C). The fruits of 18 species of *Cucurbita* were examined, and seven species contained Cucs B-D, up to 0.31% in *C. andreana,* and five contained Cucs E-I, up to 0.23% in *C. foetidissima* (Metcalf et al. 1982).

Seeds. Cucs are not usually found in the seeds of Cucurbitaceae. Rehm et al. (1957) reported that three of 45 species examined had bitter seeds, presumably from Cucs present in the surrounding tissues.

The multiplicity of Cucs and their variable distribution in the Cucurbitaceae provides a complex situation about which certain generalizations are valid: (1) Cucs B and E are the parent substances found most widely, and the other Cucs, generally present in much smaller quantities, are produced by metabolic degradation, (2) Cuc D is always associated with Cuc B, and Cuc I with Cuc E (3) Cucs G and H are always associated with Cucs B and D, (4) Cucs J and K are always associated with Cucs E and I, (5) Cuc A is always associated with Cuc B, and (6) Cuc C is singular

Table 4.2. Cucurbitacin content of fruits of *Cucurbita* spp.*

Cucurbita spp.	Cucurbitacin (mg per g of fresh weight)					
	B	D	E	I	Unknown	Glycoside
andreana Naudin	2.78	0.42				
cylindrata Bailey			0.10	0.18	trace	0.30
ecuadorensis Cutler & Whitaker	0.43	0.18				
ficifolia Bouché			-<0.02-			
foetidissima Humboldt, Bonpland & Kunth			0.36	1.59	0.49	0.91
gracilior Bailey	1.13	0.03				
lundelliana Bailey	0.63	0.15			trace	
martinezii Bailey			0.36	0.45	0.03	
maxima Duchesne			-<0.02-			
mixta Pangalo			-<0.02-			
moschata (Duchesne) Poiret			-<0.02-			
okeechobeensis (Small) Bailey			0.26	0.37	0.09	
palmata S. Watson						0.83
palmeri Bailey	0.81	0.11			0.27	
pedatifolia Bailey	0.29	0.27				
pepo Linnaeus			-<0.02-			
sororia Bailey	0.54	0.27				
texana Gray			0.07	0.37		0.75

* Data from Metcalf et al. (1982), reprinted with permission from Metcalf (1985).

and appears to occur alone in *Cucumis sativa* (Rehm et al 1957, Ferguson et al. 1983b) (7) Only Cucs B, C, and E sometimes occur alone. The distribution of Cucs in *Cucurbita* spp. is shown in Table 4.2.

At least 5 independent genes are thought to regulate the biosynthesis of Cucs (Robinson et al. 1976): a) a gene *Bi'* that regulates synthesis in seedlings, b) a gene su^{Bi} that suppresses synthesis in fruits, c) a gene that controls quantity of Cucs, d) a gene that determines the chemical nature of the Cuc formed, and e) a gene Mo^{Bi} that acts only in the presence of *Bi* and Su^{Bi} alleles to apparently control the quantity of Cuc E-glycoside (elaterinide) formation in bitter fruits (Enslin et al. 1956).

C. Behavioral Responses of Luperini Beetles to Cucurbitacins

The Cucs are arrestant and feeding stimulants (phagoincitants) for a variety of Luperini, including both Diabroticites and Aulacophorites. The adult beetles of both sexes sense the presence of the Cucs by *sensilla basiconica* on the maxillary palpi and are immediately arrested and begin to feed. Feeding on discrete amounts of Cucs in cucurbit fruit slices, dried and ground cucurbit granules, synthetic Cuc-containing starch granules,

Table 4.3. Limit of response (LR) of Diabroticina beetles to pure cucurbitacins.*

	LR in Micrograms of Cucurbitacin							
Species	B	D	E	F	G	I	L	Egly.
Diabrotica balteata	0.01			10	3	5		0.1
Diabrotica cristata	0.1		0.3					
Diabrotica barberi	0.1		0.3					
Diabrotica undecimpunctata howardi	0.001	0.03	0.01	1.0	3.0	0.1	0.01	0.05
Diabrotica undecimpunctata undecimpunctata	0.003		0.03					
Diabrotica virgifera	0.01	0.1	0.3	0.1	3.0	0.3	1.0	0.03
Acalymma vittatum	0.3		10					50

* Reprinted with permission from Metcalf (1985).

or even nanogram quantities of Cucs on silica gel thin-layer plates continues until the Cucs are consumed (Metcalf 1986). Thus the Cucs act as kairomones by promoting the host selection and feeding of the Luperini on the Cucurbitaceae. This role of the Cucs is exemplified by studies relating Cuc content of *Cucurbita* cotyledons to damage by the feeding of rootworm adults, rated on a five-point scale. For *Aulacophora foveicollis* attacking *Cucurbita moschata* cultivars, Pal et al. (1978) found a substantial correlation between Cuc content and feeding damage, (r = 0.62, n = 32, SD = 0.14) and it was concluded that "low cucurbitacin content appeared to impart resistance". For *Diabrotica u. howardi* and *Acalymma vittatum* attacking *Cucurbita pepo* cultivars, Ferguson et al. (1983b) found a strong correlation between Cuc content and feeding damage, (r = 0.77, n = 12, SD = 0.20) and concluded that there was a "strong positive correlation between cucurbitacin content and Diabroticite beetle attack". Similar correlations were found between the average numbers of Diabroticite beetles feeding on the crumpled leaves or sliced fruits of *Cucurbita* spp. and their total Cuc content (Metcalf et al. 1982); with *D. u. howardi*, leaves (r = 0.74, n = 16, SD = 0.20); and fruits (r = 0.70, n = 11, SD 0.24). With *D. v. virgifera*, the correlations were: leaves r = 0.64, n = 16, SD = 0.24; and fruits r = 0.58, n = 11, SD = 0.27.

1. Sensitivity of Diabroticite Beetles to Cucurbitacins

Qualitatively, all species of Diabroticites investigated (Table 4.3) will feed on pure crystalline Cucs A, B, C, D, E, F, G, I and L placed on filter paper or on silica gel thin-layer chromatography (TLC) plates (Chambliss & Jones 1966, Metcalf 1986).

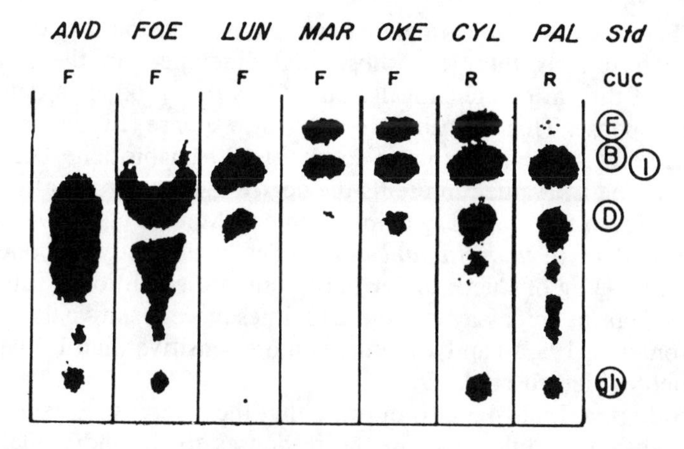

Figure 4.3. "Beetle prints" of areas from thin-layer chromatograms of *Cucurbita* fruit (F) and root (R) extracts by *D. u. howardi.* AND, *C. andreana*; FOE, *C. foetidissima*; LUN, *C. lundelliana*; MAR, *C. martinezii*; OKE, *C. okeechobensis*; CYL, *C. cylindrata*; and PAL, *C. palmata.* Reprinted with permission from Metcalf et al. 1982.

Cucurbitacins extracted from *Cucurbita* spp. by chloroform can be separated by TLC on silica gel using mixtures of solvents such as chloroform:methanol 95:5. A "beetle print" assay was developed in which the TLC plates were exposed to Diabroticite beetles so that the areas of silica gel containing Cucs were eaten away, as shown in Figure 4.3. This assay is sensitive to nanogram quantities of the various Cucs, and reveals the spectrum of Cucs present in the plant, as well as those Cucs attractive to the several species of rootworms (Metcalf 1985, 1986). The beetle feeding assay has been used to characterize the spectrum of Cucs present in roots, leaves, and fruits of 18 species of *Cucurbita* (Metcalf et al. 1980, 1982). With 5 species of Diabroticites, there were no qualitative differences between feeding patterns on a spectrum of Cucs including B, C, D, E, I and E-glycoside. Almost identical beetle prints were obtained with *A. vittatum*, which is polyphagous on Cucurbitaceae, Fabaceae, Convolvulaceae, and Poaceae, and the corn rootworms *D. barberi* and *D. v. virgifera*, which are specialists on Poaceae. *D. cristata*, whose larvae feed only on the roots of native prairie grasses, especially *Andropogon gerardi*, also fed avidly on the Cuc-containing TLC plates and produced very similar beetle prints (Metcalf 1985, 1986). *Diabrotica speciosa* and *Cerotoma arcuata* of South America have been observed to feed compulsively on the tubers of *Ceratosanthes hilariana*, a wild cucurbit containing 0.08% fresh weight of Cucs B, E, and 23,24-dihydrocucurbitacin B (Nishida et al. 1986).

When TLC plates developed from extracts of Cucurbitaceae were presented simultaneously, beetles always fed on Cuc B before Cuc E, and

Cuc D before Cuc I, indicating greater sensitivity to the B and D Cucs. Thus quantitatively there are substantial differences in the threshold amounts of the various chemically pure Cucs that produce a detectable feeding response. This quantity, the limit of response (LR), is measured on silica gel TLC plates exposed to about 100 Diabroticite beetles for four days. The LR value represents the degree of complementarity of the various Cucs to the maxillary palpi receptors (Metcalf et al. 1980). *D. u. howardi* and *D. u. undecimpunctata* beetles consistently responded by feeding on 1–3 ng of Cuc B under these standard conditions (Table 4.3). The "beetle print" bioassay is about 250 times more sensitive than HPLC (Ferguson et al. 1983b) and 1000 times more sensitive than UV spectrophotometry (Metcalf et al. 1982).

The values in Table 4.3 also indicate that the several species of beetles show quantitative differences in the responses to the individual Cucs. These differences appear to have behavioral and evolutionary significance. The conclusions drawn are: (1) Cuc B was consistently detected in the lowest amount, and is probably the parent Cuc to which Diabroticite receptors are attuned, (2) Cuc B was consistently detected at levels about 0.1 times those of Cuc E. (3) The acetoxy Cucs B and E were detected at levels about 0.1 times those of the corresponding desacetoxy Cucs D and I respectively. (4) Saturation of the desacetoxy Cucs at the $C_{23} = C_{24}$ bond (Cuc L) had little effect on the LR values. (5) Sensitivity to the 2-OH, 3-C = O Cuc D was greater than that to the 2-OH, 3-OH Cuc F. (6) The *fucata* group of species *D. balteata*, *D. u. howardi* and *D. u. undecimpunctata*, are more sensitive to the phagostimulant Cucs than the *virgifera* group species *D. barberi*, *D. cristata*, and *D. v. virgifera*. (7) *A. vittatum* is substantially less sensitive to Cucs than the *Diabrotica* spp (Metcalf et al. 1982).

The singular arrestant and phagostimulant effects of Cucs upon Luperini beetles also occurs with the Old World *Aulacophora* spp. A. *foveicollis* has been shown to feed on pure Cuc E on filter paper (Sinha & Krishna 1970), and *A. coffeae*, *A. femoralis*, and *A. nigripennis* are phagostimulated by Cuc B and other Cucs (Nishida & Fukami 1990).

Based on the presence of cucurbitacin arrestant and phagostimulant responses in a considerable range of Luperini from both the Old World and the New World (Figure 4.1), we have suggested that the entire group of Luperini rootworm beetles evolved from a common ancestor that coevolved with the Cucurbitaceae, and that present day preferences for other hosts, such as grasses, are relatively recent (Metcalf 1985, 1986).

D. Cucurbitacins as Allomones for Diabroticites

There is no ambiguity about the role of cucurbitacins as protective allomones against herbivore attack. Cucs B and E and their derivatives are the bitterest substances known, and can be detected by humans at dilutions as great as 1 ppb (Metcalf et al. 1980). Ingestion of trace amounts, e.g. by tasting the cotyledons of *C. maxima* or an elytron of a *D. u. howardi* collected from a bitter plant, produces an almost paralytic response on lips and mouth and a persistent after-taste. The Cucs are extremely toxic to mammals, with LD_{50} values to mice intraperitoneally of Cuc A 1.2, Cuc B 1.1 and Cuc C 6.8 mg per kg (David & Vallance 1955); and orally of Cuc I 5.0 and Cuc E-glycoside of 40 mg per kg (Stroesand et al. 1985). Cattle and sheep feeding upon bitter Cucurbitaceae during drought conditions have been poisoned (Watt & Breyer-Brandwyk 1962) and outbreaks of human poisoning have resulted from ingestion of fruits of *Cucurbita* cultivars that have reverted to the heterzygous *Bi* alleles for bitterness (Ferguson et al. 1983a and Rhymal et al. 1984).

The Cucs are feeding deterrents for many species of insect herbivores, including the leaf beetles *Phyllotreta* spp., *Phaedon* spp. and *Ceratoma trifurcata*, the stem borer *Margonia hyalinata*, and red spider mites (DaCosta & Jones 1971b, Metcalf et al. 1980, and Nielson et al. 1977).

1. Storage and Sequestration of Cucurbitacins by Luperini Beetles

The incredibly bitter taste of Cucs and their toxic effects on vertebrates suggest a further role as protective allomones of Diabroticites against predators, including birds. Luperini beetles can concentrate and sequester relatively large amounts of Cucs in free and derivatized form. Groups of *D. balteata* fed on bitter *Cucurbita* fruits for varying periods of time contained a Cuc conjugate in the hemolymph that increased from about 1 μg per μl after 1 day to 22μg per μl and reached a maximum of 25–32 μg per μl after 14–28 days (Ferguson et al. 1985). Similar Cuc blood levels of 20–26 μg per μl were detected in *D. u. howardi* adults collected from bitter squash plants. This Cuc content is equivalent to about 15 mg Cucs per g of body weight. No Cucs were detectable in *D. balteata* adults reared on pollen food (Ferguson et al. 1985). Andersen et al. (1988) showed that *D. u. howardi* and *D. v. virgifera* adults metabolize Cuc D to the C_{23}-C_{24} dihydro *O*-glucoside. Two Brazilian leaf beetles, *D. speciosa* and *Ceratoma arcuata* were observed feeding voraciously on the roots of *Ceratosanthes hilariana*, a wild cucurbit containing about 0.08% fresh weight of Cucs B and D and 23,24-dihydrocucurbitacin D (Nishida et al. 1986). The tissues of these Diabroticites contained 23,24-dihydrocucurbitacin D to levels of about 6μg per insect for *D. speciosa*, and 20μg per insect for *C. arcuata* (Nishida & Fukami 1990).

From the viewpoint of evolutionary zoogeography, it is particularly interesting that both New World and Old World Luperini leaf beetles respond to Cuc phagostimulants and selectively sequester the same metabolite, 23,24-dihydrocucurbitacin D (Ferguson et al. 1985, Andersen et al. 1988, Metcalf 1986, Nishida & Fukami 1990).

2. *Effects of Cucurbitacin Sequestration upon Predators*

Body levels of Cucs in Diabroticites fed upon bitter *Cucurbita* fruit are sufficiently high to deter predation of these beetles by the Chinese mantis *Tenodera aridifolia sinensis*. When such mantids were offered four species of adult Diabroticites fed a pollen diet, none of the beetles were ever rejected nor did their consumption elicit any atypical behavior (Ferguson & Metcalf 1985). However, when offered adult beetles fed upon bitter *Cucurbita* fruit, the rejection rates were *D. balteata* 72%, *D. u. howardi* 48%, and *D. v. virgifera* 24%. All of these rejection rates were significant at $P < 0.005$–0.05. When pollen fed *D. u. howardi* adults were topically treated with Cuc B in acetone at 14μg per beetle, these beetles were consistently rejected by the Chinese mantids (Ferguson & Metcalf 1985). Rejection of the bitter beetles consisted of the mantid immediately flinging away the beetle in < 10 seconds after a bite on one elytron, followed by a period of marked unsteadiness, regurgitation, and/or excessive grooming. In some instances, the mantid fell from its perch and occasionally, after long periods of moving about and holding the bitter beetle away from its body, the predator would retaste the beetle before discarding it. The hemolymph of the *D. balteata* beetles used in these experiments averaged 22 μg Cucs per μl (Ferguson & Metcalf 1985).

The hemolymph Cuc constituents were very persistent, and *D. u. howardi* fed on bitter *Cucurbita* fruit and subsequently fed on a pollen diet for 6 weeks still contained quantities of Cucs comparable to those in beetles fed the bitter fruit continuously. After an initial 3 week feeding period on bitter *Cucurbita* fruit, followed by feeding on a pollen diet for nearly three months, 100% of the *D. u. howardi* beetles were rejected by the mantids (Ferguson & Metcalf 1985). This demonstrated that the ingested and metabolized Cucs are long term storage products. Cucs, probably hemolymph derived, were found in the eggs of *D. balteata* and *D. u. howardi* that were fed on bitter fruit as adults. This is shown in the beetle prints of Figure 4.4.

The striped cucumber beetle, *Acalymma vittatum*, presented a special case. There was not a significant rejection rate between adults fed on fruit

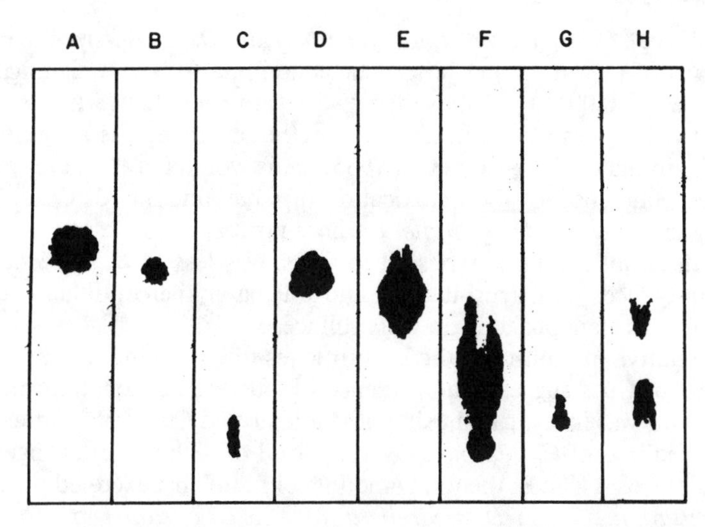

Figure 4.4. "Beetle prints" of areas eaten from thin-layer chromatograms by *D. u. howardi*. A, Cuc B; B, Cuc D; C, hemolymph from field collected *D. howardi*; D, hemolymph incubated with pectinase prior to TLC; E, excreta of *D. howardi* fed Cuc containing fruit; F, body of *D. howardi* with no exposure Cucs for 6 weeks; G, extract of 200 *D. howardi* eggs; H, extract of 400 eggs of *D. howardi* exposed to Cuc only as larvae. Reprinted with permission from Ferguson et al. 1985.

containing Cucs and those fed a pollen diet. However, the *A. vitattum* adults which were reared as larvae on the roots of the Cuc-containing *Cucurbita maxima* produced eggs containing Cucs (Ferguson and Metcalf 1985, Metcalf 1986). Therefore, both the pollen fed and bitter fruit fed *A. vittatum* contained appreciable amounts of Cucs and the rejection rates by the mantids were expected to be comparable. It is apparent that larvae fed on bitter Cuc-containing roots sequestered Cucs that were transferred to the adults during metamorphosis. The sequestration of Cucs in the eggs of Diabroticite beetles suggests that these "bitter" eggs act as allomones to deter ant predators such as *Solenopsis* and *Pheidole* (Ferguson & Metcalf 1985). Cucurbitacins have also been shown to be feeding deterrents to the Japanese tree sparrow *Passer montunus saturatus* (Nishida & Fukami 1990).

3. Disposition and Fate of Cucurbitacins in Diabroticites

In contrast to the high toxicity of Cucs to vertebrates, the Diabroticite beetles are relatively immune, and when fed Cucs at > 2000 mg per kg, *D. u. howardi* and *D. v. virgifera* showed no appreciable acute effects

(Metcalf et al. 1980). However, for *D. barberi* and *D. v. virgifera*, the mean life spans of the adults fed bitter Cuc-containing fruit was significantly shorter ($P < 0.001$) than those fed sweet fruit. For *A. vitattum* no significant differences were detected in females, but there was a significant decrease in males (Ferguson et al. 1985). This would appear to represent the metabolic costs of the detoxication of large amounts of Cucs. This stress was greatest in *D. v. virgifera*, whose normal life-style as a Poaceae feeder does not regularly expose it to Cucs, was less in *D. balteata* that commonly feeds on Cucurbitaceae, and was barely perceptible in *A. vittatum* that is stenophagous on Cucurbitaceae.

The relative immunity of the Diabroticites following ingestion of large amounts of Cucs suggests the presence of efficient detoxication mechanisms. This was investigated using ^{14}C-radiolabeled Cuc B biosynthesized in the seedlings of *C. maxima*. Over a period of 48 hours after ingesting the radiolabeled Cuc B, the total amounts of radiolabel excreted was 67% in *A. vittatum*, 77% in *D. v. virgifera*, 81% in *D. cristata*, 85% in *D. u. howardi*, and 95% in *D. balteata*. Cuc B was identified as a minor component of the radiolabeled products in the excreta, but the preponderance of the ^{14}C was present in the excreta as three polar metabolites which comprised 47% of the total ^{14}C excreted in *D. v. virgifera*, and 91% in *D. balteata*. The major excretory metabolite in the five Diabroticites examined was a Cuc glycoside conjugate that comprised from 25% of the total ^{14}C excreted in *A. vittatum* to 44% in *D. cristata* (Ferguson et al. 1985). The principal excretory metabolite of *D. u. howardi* and *D. v. virgifera* fed Cuc B was identified as 23,24-dihydro Cuc D glucoside (Andersen et al. 1988).

E. Coevolutionary Implications of Cucurbitaceae/ Luperini Association

The original evolutionary strategy of the Cucurbitaceae to synthesize the intensely bitter and toxic Cucs as allomones was very successful, as major herbivores are relatively few. This strategy has produced three major variations in the structure of the Cuc allomones: (1) C_1-C_2 unsaturation, (2) acetylation at C_{23}, and (3) glucosylation (Metcalf et al. 1982).

Comparisons of the phylogeny of Cucurbitaceae, as constructed from genetic compatibilities and numerical taxonomy, with Cuc content suggests that the original primitive form of Cucs was B-D and that Cucs E-I and the glycosides were secondary modifications, that each evolved at least twice independently in *Cucurbita*, i.e. $C_1 = C_2$ (E-I Table 4.2) in the *C. martinezii* and *C. palmata* sub-groups, and formation of glycosides in *C. texana*, *C. palmata*, and *C. cylindrata* (Metcalf et al. 1980, 1982).

All species of Diabroticites examined have specific Cuc receptors on the maxillary palpi attuned to Cuc B, and the presence of trace quantities elicits arrest and phagostimulation in these species whether they are presently stenophagous on Cucurbitaceae, as *A. vittatum*, oligophagous on Cucurbitaceae as in *D. balteata*, polyphagous as *D. u. howardi* and *D. speciosa*, oligophagous on Poaceae, as *D. barberi* and *D. v. virgifera*, or monophagous on *Andropogon*, as *D. cristata*. This behavior appears to extend to a wide range of Old World *Aulacophora* spp. (Nishida & Fukami 1990).

All species of Diabroticites examined have well developed detoxication and storage mechanisms to partially metabolize, conjugate, and sequester Cucs. Thus they are immune to the generally repellent and acutely toxic effects of Cucs (Ferguson et al. 1985).

Natural enemies of Diabroticites are few, and the coevolutionary strategy to sequester Cucs as allomones against predation provides a successful survival mechanism.

F. Coevolutionary Scenario

The coevolutionary interactions between Cucurbitaceae and Luperini beetles can be portrayed as (DaCosta & Jones 1971a, Metcalf 1979, 1986, Price 1984):

(1) ancestral Cucurbitaceae with *bibi* genes for Cuc synthesis are heavily fed upon by herbivores.

(2) mutation in Cucurbitaceae to *Bibi* forms bitter and toxic Cucs that deter herbivore attack.

(3) strong selection pressures spread *Bibi* genes throughout evolving Cucurbitaceae species.

(4) mutant Cucurbitaceae flourish in the absence of herbivore attacks.

(5) mutant ancestral Luperini beetle develops detoxication and excretory pathways to reduce harmful effects of Cucs.

(6) Luperini beetles expand into new ecological niches developing specific receptors to detect Cucs.

(7) Luperini beetles develop high blood and tissues levels of Cuc conjugates for defense against predators.

IV. ATTRACTION OF DIABROTICITES TO CUCURBITA BLOSSOMS

The blossoms of Cucurbitaceae are well known to be highly attractive to Luperini beetles. Historically the northern corn rootworm *D. longicornis* (= *barberi*) was first described from specimens collected from the blos-

Table 4.4. *Diabrotica* spp. Beetle Populations in *Cucurbita* spp. Blossoms, Central Illinois, Aug.–Sept. 1977*

Cucurbita sp.	No. Blossoms	Beetles per blossom (mean ± S.D.) *D. u. howardi*	*D. v. virgifera*
andreana	270	16.1 ± 8.0	6.0 ± 4.9
gracilior	185	5.8 ± 3.1	3.0 ± 4.2
lundelliana	150	8.0 ± 4.9	2.0 ± 3.2
martinezii	112	7.3 ± 1.6	0.6 ± 0.6
maxima	320	11.0 ± 5.9	11.6 ± 8.0
mixta	194	2.6 ± 1.6	4.1 ± 3.6
moschata	300	5.2 ± 2.0	6.1 ± 4.1
okeechobeensis	90	12.0 ± 4.8	0.6 ± 0.8
palmeri	153	4.8 ± 1.2	1.8 ± 2.0
pedatifolia	5	11.4 ± 8.6	6.9 ± 13.4
pepo	75	2.0 ± 1.7	13.1 ± 5.5
sororia	58	6.6 ± 6.1	1.7 ± 2.3
texana	253	5.0 ± 2.2	4.3 ± 4.5

* Metcalf & Rhodes, unpublished data.

soms of the buffalo gourd *Cucurbita foetidissima* in 1824 and the western corn rootworm was first collected from blossoms of the same plant in 1868 (Smith & Lawrence 1967). Luperini beetles of both *Aulacophorina* (Takizawa 1978) and the *Diabroticina* (Krysan et al. 1984) are frequently collected from the blossoms of wild and cultivated Cucurbitaceae. The various Diabroticite spp. exhibit distinct preferences for the male blossoms of cultivated *Cucurbita. D. barberi, D. u. howardi* and *Acalymma vittatum* were found in greater abundance in the flowers of *C. maxima* than in those of *C. pepo* (Fronk & Slater 1956). The order of preference by *D. barberi* and *D. v. virgifera* among interplantings of *Cucurbita* spp. was *C. maxima* > *C. pepo* > *C. mixta* > *C. moschata* (Fischer et al. 1984). However, *D. barberi* was found almost exclusively in the flowers of *C. maxima*, whereas *D. v. virgifera* was present in both *C. maxima* and *C. pepo* flowers. In interplantings of *Cucurbita* spp. *D. u. howardi* and *D. v. virgifera* preferred *C. maxima* flowers over those of *C. moschata* and *C. pepo* (Andersen & Metcalf 1987).

During Aug.-Sept. 1977, blossoms from 13 *Cucurbita* species planted in replicated blocks were routinely sampled twice weekly for populations of *D. u. howardi* and *D. v. virgifera* as shown in Table 4.4. The data show that the various *Cucurbita* spp. blossoms differed substantially in attraction for the rootworm adults and that the relative attraction of the two *Diabrotica* species was different, e.g. the blossoms of *C. martinezii* and *C. okeechobeensis* were strongly preferred by *D. u. howardi* as compared to *D. v. virgifera*, and the latter species strongly preferred *C. pepo* blossoms as compared to *D. u. howardi* (Table 4.4) (Metcalf & Rhodes, unpublished data).

The aggregation of Diabroticite beetles in *Cucurbita* blossom is the result of the concerted attraction from an array of blossom volatiles that increases the arrival rate of the beetles, acting together with the arrestant and phagostimulant properties of cucurbitacins present in the blossoms that delay departure (Andersen & Metcalf 1987). The blossoms of the highly attractive *C. maxima* (Table 4.4) contained 137–147 μg per blossom of Cuc B and traces of Cuc D (Andersen & Metcalf 1987). The high Cuc content of the blossoms of *Cucurbita andreana* and *C. maxima* and hybrids resulted in such a high rate of destruction by Diabroticite beetles that fruit set was severely impaired (Rhodes et al. 1980, Metcalf & Rhodes 1990).

A. Long Range Attraction of Diabroticites to *Cucurbita* Blossoms

Specific proof of the role of blossom volatiles in the attraction of Diabroticites was obtained by placing 30 g of shredded blossoms in 1 l paper cartons coated with insect adhesive and covered with cheesecloth. After 1 hour in a cucurbit plot, the average number of beetles caught on four blossom baited sticky traps was (Metcalf & Lampman 1989a):

	C. maxima	*C. moschata*	Unbaited
D. u. howardi	3.0 ± 1.4	0.5 ± 0.6	0.5 ± 0.6
D. v. virgifera	86.6 ± 30.6	11.3 ± 6.2	6.8 ± 7.5
A. vittatum	11.8 ± 4.5	2.8 ± 1.5	1.8 ± 0.5

These data clearly demonstrated the role of blossom volatiles as lures for Diabroticites, and showed that *C. maxima* blossom volatiles were much more attractive than those of *C. moschata.*

B. Chemical Identification of Attractive *Cucurbita* Blossom Volatiles

Fractionation of volatiles collected from the blossoms of *Cucurbita* cultivars showed the presence of numerous components: *C. maxima* ca. 40, *C. moschata* ca. 16, and *C. pepo* ca. 12. A number of these were identified by GC-MS as shown in Table 4.5 (Andersen & Metcalf 1987). Indole was specifically identified as a blossom component of *C. maxima* that was attractive to both *D. v. virgifera* and *A. vittatum*, but not to *D. u. howardi.* Rootworm beetle preference for *C. maxima* flowers was correlated with high release rates of volatiles such as indole, cinnamaldehyde, cinnamyl

Table 4.5. *Cucurbita maxima* blossom volatiles and the limits of response (LR) by *Diabrotica* spp. on sticky traps.

	LR (mg)			
Blossom volatile	*D. barberi*	*D. cristata*	*D. u. howardi*	*D. v. virgifera*
1,4-dimethoxybenzene	>100	>100	>100	>100
1,2,4-trimethoxybenzene	>100	>100	100	>100
benzyl alcohol	>100	>100	>100	>100
benzaldehyde	100	100	100	100
phenethanol	10–30		100	>100
phenylacetaldehyde	>100		10–30	100
4-methoxybenzyl alcohol	>100		>100	>100
4-methoxybenzaldehyde	>100		>100	>100
indole	>100	100	100	1.0
cinnamyl alcohol	1.0	3–10	10–30	100
cinnamaldehyde	100	10–30	0.3	30–100
α-ionone	>100	>100	>100	>100
β-ionone	100	100	100	1.0
nerolidol	>100		>100	>100
4-methoxycinnamaldehyde*	100	30–100	>100	0.03
4-methoxyphenethanol*	0.1	1–3	>100	100

* not identified as a blossom volatile.

alcohol, and β-ionone, and with the presence of Cucs (Andersen & Metcalf 1986).

In a comprehensive investigation of the blossom volatiles of *C. maxima* cultivars, Andersen (1987) isolated 31 components from the steam distillate and identified 28 specific chemical structures by GC-MS. Using sticky, cylindrical paper cartons (Table 4.5) more than 20 of these volatile blossom components have been evaluated in the field as attractants for the several species of Diabroticite beetles. (Andersen & Metcalf 1986, Lampman et al. 1987, Lampman & Metcalf 1987, 1988, Metcalf & Lampman 1989a,b, Lewis et al. 1990). A summary of the results is presented in Table 4.5.

The *Cucurbita* blossom volatiles include the "green volatiles": six carbon alcohols and aldehydes, phenyl propanoids, and the terpenoid β-ionone. The phenyl propanoids, indole, cinnamaldehyde, and cinnamyl alcohol are important volatile attractants, singly or in combination, for the several species of Diabroticites, and have LR values in field trapping of 1–3 mg for at least one of the four *Diabrotica* spp. investigated. Phenyl propanoids are $C_6 \cdot C_3$ compounds formed in the shikimic acid pathway through phenylalanine to cinnamic acid (Friedrich 1976). They are produced in relatively large amounts during pollen maturation in blossoms (Wierman 1970, 1981). Thus it appears that the remarkable long range attraction of many Diabroticite spp. to the blossoms of Cucurbitaceae implies that the original coevolutionary association between Cucurbitaceae and Luperini beetles was that of pollen seeking and consequent fertilization. Coleoptera are thought to be the most primitive pollenizers of angiosperms, and were probably associated with open bowl-shaped flowers (Kevan & Baker 1983).

V. DEVELOPMENT OF LURES FOR DIABROTICITES

The first report of volatile attractants for Diabroticites was that of Morgan & Crumb (1928), who observed *D. u. howardi* beetles to be attracted to cinnamaldehyde and cinnamyl alcohol. Ladd et al. (1983) found that eugenol was an attractant for *D. barberi* beetles, and Ladd (1984) showed that eugenol, isoeugenol, and the structurally related 2-methoxy-4-propylphenol were effective attractants for this species. Indole was isolated from the blossoms of *Cucurbita maxima* and shown to be an attractant for *D. v. virgifera* and *Acalymma vittatum*, but was not attractive to *D. u. howardi* (Andersen & Metcalf 1986). A variety of other aromatic compounds, including phenylacetaldehyde, benzyl acetone, phenethyl alcohol, benzyl alcohol, and veratrol were found to be moderately attractive to *D. u. howardi* adults (Lampman et al. 1987). *D. cristata*, closely related to *D. barberi*, was found to share the latter's attraction to eugenol and 2-methoxy-4-propylphenol (Yaro et al. 1987).

Estragole, or 4-methoxy-1-allylbenzene, closely related to eugenol, was shown to be an attractant for *D. v. virgifera*, but not for *D. barberi* or *D. u. howardi* adults (Metcalf & Lampman 1989a), and cinnamyl alcohol was found to be an effective attractant for *D. barberi*, but not for *D. v. virgifera* or *D. u. howardi* adults (Metcalf & Lampman 1989c).

A. Comparative Attractant Responses of Diabroticites to Eugenol and Estragole

These two phenylpropanoids are closely related although their specific attractivity to individual Diabroticite species remains intriguing. As shown in Table 4.6 (Lampman et al. 1987), eugenol is specifically attractive to *D. barberi*, and estragole to *D. v. virgifera*. Neither compound was attractive to *D. u. howardi*, although chavicol (4-hydroxy-1-allylbenzene) was somewhat attractive to this species. The closely related phenylpropanoids methyl eugenol (3,4-dimethoxy-1-allylbenzene), safrole (3,4-methylenedioxy)-1-allylbenzene, anethole (4-methoxy-1-propenylbenzene), isoeugenol, and methyl isoeugenol were either unattractive or of very low attractivity to all three species of corn rootworm adults. This

Table 4.6. Attractivity of Eugenol, Estragole, and Derivatives to Adult Corn Rootworms.*

Lure	1 day sticky trap catch—100 mg (mean ± S.D.)[1] *D. barberi*	*D. u. howardi*	*D. v. virgifera*
eugenol	6.2 ± 1.6b	4.7 ± 11.ab	2.2 ± 1.0a
isoeugenol	3.7 ± 0.7b	5.2 ± 2.2ab	3.2 ± 0.5ac
chavicol	0.7 ± 0.2a	10.2 ± 4.1b	3.5 ± 1.0ac
methyl eugenol	0.7 ± 0.2a	4.0 ± 1.7ab	2.2 ± 1.0a
safrole	0.5 ± 0.5a	2.2 ± 0.5a	1.2 ± 1.2a
estragole	0.2 ± 0.2a	6.5 ± 3.3ab	23.5 ± 5.9b
anethole	1.0 ± 0.4a	2.2 ± 0.8a	7.7 ± 1.8c
control	0a	1.5 ± 0.5a	3.5 ± 0.6ab

[1] means followed by different letters are significantly different ($P \leq 0.05$).

* Data from Lampman et al. (1987).

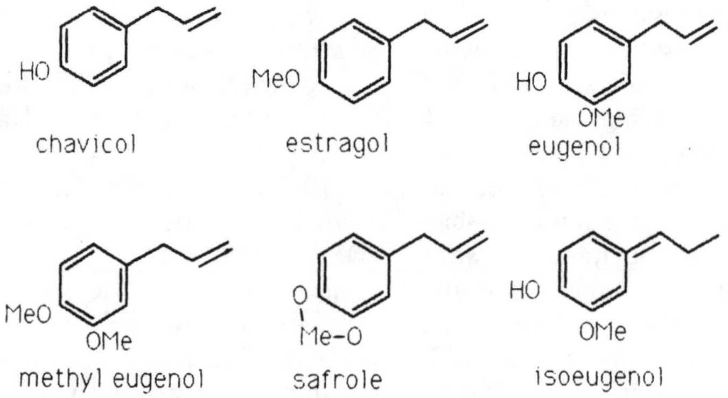

HO chavicol

MeO estragol

HO OMe eugenol

MeO OMe methyl eugenol

O Me-O safrole

HO OMe isoeugenol

data (Table 4.6) shows that the attractivity of these phenylpropanoid compounds is highly specific, depending upon the nature and position of the oxygen-containing substituents of the aryl ring. Phenylpropanoids with allyl ($-CH_2CH = CH_2$) side chains were generally more attractive than those with propenyl ($-CH = CHCH_3$) side chains, e.g. estragole was more attractive to *D. v. virgifera* than anethole. Saturation of the side chain, as in 4-methoxy-1-propylbenzene, decreased attractivity to *D. v. virgifera.* In a continuation of these investigations of lure specificity (Lampman & Metcalf 1988), eugenol and isoeugenol were shown to be about equally attractive to *D. barberi* and to *D. cristata.* Cinnamyl alcohol was also highly attractive to these two species, but was not attractive to *D. v. virgifera* or to *D. u. howardi. D. cristata* also shares some chemosensory adaptations with *D. v. virgifera*, as both species were attracted to β-ionone, a terpenoid present in the blossoms of *Cucurbita maxima*, but

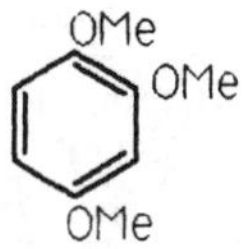

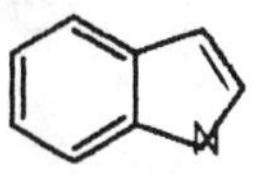

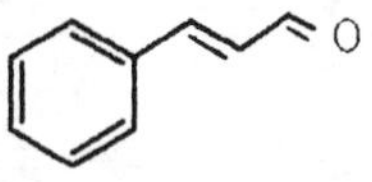

1,2,4-trimethoxybenzene indole *trans*-cinnamaldehyde

neither *D. barberi* nor *D. u. howardi* were attracted. Additional investigations have emphasized these evolutionary lure affinities in that 4-methoxycinnamaldehyde is highly attractive to *D. cristata* and *D. v. virgifera*, but is unattractive to *D. barberi* or *D. u. howardi* (Lampman & Metcalf 1988), and 4-methoxyphenethanol is highly attractive to both *D. barberi* and *D. cristata*, but not to *D. u. howardi* or *D. v. virgifera* (Metcalf & Lampman, unpublished).

In summary, each of the four Diabroticites examined displays a distinctive pattern of lure responses when exposed to a broad spectrum of kairomone-type lures. It is evident, however, that the three closely related species of the *virgifera* subgroup share chemosensory affinities with one another, and that the responses of *D. cristata* represent an intermediary form between *D. barberi* and *D. v. virgifera*.

The evolutionary basis for Diabroticite attraction to the eugenol-type secondary plant compounds is unclear. These phenylpropanoids have not yet been identified in either *Cucurbita* blossoms or in corn silk. Yaro et al. (1987) conjectured that the adaptation of *D. cristata* and *D. barberi* to eugenol and isoeugenol may have arisen in a Nearctic ancestor as an evolutionary adaptation to adult feeding on prairie phorbs. However, the attraction of *D. u. howardi* of the *fucata* subgroup to various phenylpropanoids that are also attractive to the *virgifera* spp. (Lampman & Metcalf 1988, Lampman et al. 1987) suggests that the chemosensory responses to phenylpropanoids are of older evolutionary origin, perhaps parallel to the generalized responses of the Diabroticites to cucurbitacins (Metcalf & Lampman 1989a).

B. Synergism of *Cucurbita* Blossom Volatiles as Attractants

The aroma of the blossoms of the Cucurbitaceae is obviously attractive to a wide range of Luperini beetles found infesting the blossoms. The chemical characterization of aroma constituents from *Cucurbita maxima* (Andersen 1987, Andersen & Metcalf 1987), and quantitative field bioassays of specific blossom aroma constituents (Andersen & Metcalf 1986, Lampman et al. 1987) provided the rationale for the development of

Table 4.7. Attractivity of Blossom Volatile Mixtures to Adult Corn Rootworms*

Mixture	Dose (mg)	1 day sticky trap catch (mean ± S.D.)[1] *D. u. howardi*	*D. v. virgifera*
veratrol, indole, phenylacetaldehyde (VIP)	0	5.2±4.8a	2.7±1.6a
	1	6.2±2.9a	4.2±4.8a
	3	15.7±8.2b	4.7±2.6a
	10	37.7±22.6b	12.5±3.3b
	30	72.5±20.2c	15.5±2.6b
trimethoxybenzene, indole, cinnamaldehyde (TIC)	0	5.2±4.8a	2.7±1.6a
	1	15.2±11.9ab	15.5±2.5b
	3	17.0±8.5b	36.0±16.9c
	10	59.9±29.7c	57.0±16.2cd
	30	76.0±21.4c	77.2±21.3d

[1] means followed by different letters are significantly different ($P \leq 0.05$).

* Data from Lampman & Metcalf (1987).

kairomone lures for Diabroticite beetles. Indole was shown to be a specific attractant for *D. v. virgifera* and *A. vittatum* (Andersen & Metcalf 1986), but was unattractive to *D. u. howardi* and *D. barberi*. Phenyl acetaldehyde present in *Cucurbita* blossom volatiles was found to be attractive to *D. u. howardi*, but not to the other species. Structure-activity studies on these and related compounds showed that *ortho*-dimethoxybenzene (veratrole) was much more attractive to *D. u. howardi* than the *meta*- and *para*-isomers (Lampman & Metcalf 1987). Various mixtures of these three compounds showed a marked degree of olfactory synergism to *D. u. howardi*, and equal proportions of veratrole, indole and phenylacetaldehyde (VIP mixture) was highly attractive to *D. u. howardi*, but unattractive to *D. v. virgifera* (Lampman and Metcalf 1987) (Table 4.7).

The presence of cinnamaldehyde and 1,2,4-trimethoxybenzene in *C. maxima* blossoms and their structural analogies to phenylacetaldehyde and veratrole led to the improved volatile mixture of trimethoxybenzene-indole-cinnamaldehyde (TIC mixture) (Table 4.7) that not only was found to be attractive to a variety of Diabroticites, but also shared olfactory synergism to all species investigated (Lampman & Metcalf 1987, 1988, Lampman et al. 1991). The TIC mixture was highly attractive to *D. u. howardi* and *D. v. virgifera* in field studies with cylindrical sticky traps over a dosage range of 1–30 mg (Table 4.7). During 1986–1989, field bioassays of the TIC mixture and its individual components on cylindrical sticky traps in corn fields showed that TIC was 2.2 ($n = 7$) times more active than the expected additive response for *D. v. virgifera*, 2.1 times

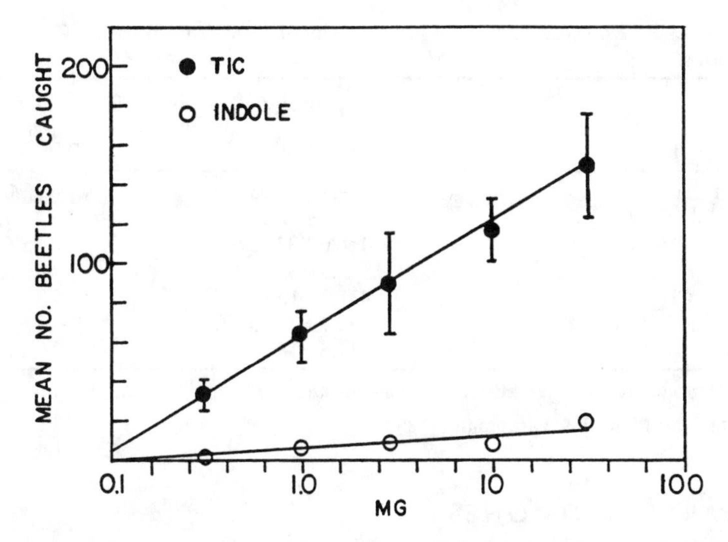

Figure 4.5. *Acalymma vittatum* adults caught on cylindrical sticky traps baited with a range of doses of indole and with TIC mixture (trimethoxybenzene, indole, and cinnamaldehyde). Reprinted with permission from Lewis et al. 1990.

more active for *D. barberi* (a single test) and 1.3 times more active for *D. u. howardi* (n = 4), (Lampman et al. 1991). The olfactory synergism was lowest for *D. u. howardi* beetles because of their strong response to cinnamaldehyde.

In field tests with cylindrical sticky traps in a field of *Cucurbita moschata*, the TIC mixture was found to be highly attractive to the striped cucumber beetle *Acalymma vittatum* (Lewis et al. 1990). As shown in Figure 4.5, plots of log dose per trap vs. mean number of adults captured were linear for both indole and the TIC mixture, but the slope of the log dose response curve for the TIC baited traps was 7 times greater than that for indole. The traps baited with the three-component TIC mixture caught 2.0 times as many *A. vittatum* adults as the theoretical additive response of the components, again demonstrating olfactory synergism (Lewis et al. 1990).

We consider the TIC mixture to be a highly simplified cucurbit blossom volatile aroma. The addition of other blossom volatiles such as β-ionone did not improve the efficacy, and the two component mixture of cinnamaldehyde and indole, although highly attractive, produced only additive attractivity (Metcalf & Lampman 1989a). The substitution of cinnamyl alcohol for cinnamaldehyde in the mixture resulted in decreased attractivity for *D. v. virgifera* adults and increased attraction for *D. barberi* adults (Metcalf & Lampman unpublished).

Table 4.8. Role of Methoxy Group in Selectivity of Attractants for Adult Corn Rootworms*

Lure	1 day sticky trap catch—100 mg (mean ± S.D.)[1] *D. u. howardi*	*D. v. virgifera*
$C_6H_5CH_2CH{=}CH_2$	24.7 ± 16.9a	1.0 ± 0.8a
$4\text{-}CH_3OC_6H_4CH_2CH{=}CH_2$	9.2 ± 5.6a	17.2 ± 7.8c
$C_6H_5CH{=}CHCN$	156.0 ± 37.2c	1.5 ± 6.6ab
$4\text{-}CH_3OC_6H_4CH{=}CHCN$	6.5 ± 3.0a	59.0 ± 32.1d
$C_6H_5CH{=}CHC(O)H$	431.2 ± 120.3d	4.0 ± 1.8b
$4\text{-}CH_3OC_6H_4CH{=}CHC(O)H$	42.7 ± 27.6b	143.7 ± 75.3c
control	17.5 ± 18.3a	0.7 ± 0.9a

[1] means followed by different letters are significantly different ($P \leq 0.05$).

* Data from Metcalf & Lampman (1989a).

VI. PARAKAIROMONES

The demonstrations of the effectiveness of cinnamaldehyde as a lure for *D. u. howardi*, and of estragole as a lure for *D. v. virgifera*, focused attention on the role of small changes in the molecular structure of phenylpropanoids on kairomone-type lures for Diabroticites. Specifically, it became apparent that there was an almost "all-or-none" responsiveness among the several *Diabrotica* spp. associated with changes in the substituents of the phenyl rings, as well as with terminal groups of the unsaturated side chains (Table 4.6).

To explore this, field evaluations were made of the attractivity of three pairs of phenylpropanoid attractants, each pair with a different side chain ($-CH_2CH = CH_2$, $-CH = CHC(O)H$, or $-CH = CHCN$); and one of each pair with a *para*-CH_3O-phenyl substitution. The results of attractivity studies on *D. u. howardi* and *D. v. virgifera* adults were dramatic. It is evident from the data in Table 4.8 that the captured species ratio (*howardi/virgifera*) shifts by more than a factor of 10 with the *para*-CH_3O-substitution of cinnamaldehyde and cinnamonitrile. In essence, cinnamaldehyde and its bioisostere cinnamonitrile are excellent lures for *D. u. howardi* and unattractive to *D. v. virgifera*, and conversely 4-methoxycinnamaldehyde and its bioisostere 4-methoxycinnamonitrile are excellent lures for *D. v. virgifera* and unattractive to *D. u. howardi*.

The 4-methoxycinnamaldehyde is the most effective lure yet discovered for *D. v. virgifera* and is also attractive to *D. cristata*, but is unattractive to *D. barberi* (Lampman & Metcalf 1988). Significant attraction of *D. v. virgifera* in field tests was obtained with as little as 30 μg of 4-methoxycinnamaldehyde on wicks of cylindrical sticky traps (Figure 4.6), and wicks treated with 100 mg remained more than 50 times more attractive

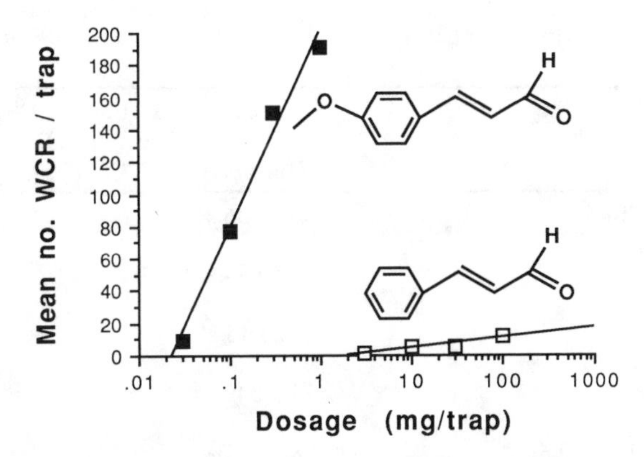

Figure 4.6. Attraction of *D. v. virgifera* (WCR) adults to cylindrical sticky traps baited with varying doses of cinnamaldehyde and 4-methoxycinnamaldehyde. Reprinted with permission from Metcalf & Lampman 1991.

than controls over an 18-day period of exposure (Metcalf & Lampman 1989b). The 4-methoxycinnamaldehyde has been identified as a constituent of a variety of plants, including *Agastache rugosa*, *Orthodon methylchavicoliferum*, and *Ocimum basilicum* (Lamiaceae); *Artemesia dracunculus* and *Sphaeranthus indicus* (Asteraceae); *Acorus gramineus* (Araceae); and *Limnophila rugosa* (Scrophulariaceae) (Metcalf & Lampman 1989b). The presence of the 4-methoxycinnamaldehyde in these essential oils was consistently associated with much larger quantities of estragole which is evidently a precursor. The 4-methoxycinnamonitrile is not a natural product, and its high attractivity to *D. v. virgifera*, like that of cinnamonitrile to *D. u. howardi*, is associated with the biochemical equivalence of the aldehyde $C = O$ and nitrile $C \equiv N$.

The demonstration of the pronounced attractivity of cinnamyl alcohol to *D. barberi* adults, but not to those of *D. v. virgifera* and *D. u. howardi* (Metcalf & Lampman 1989c) again focused attention on structural manipulation of the basic phenylpropanoids. Attractivity to *D. barberi* was found to be related to a terminal-CH_2OH group on the phenylpropanoid side chain, and 3-phenyl-1-propanol (phenpropanol), a bioisostere of cinnamyl alcohol, was shown to be nearly as effective as a lure for *D. barberi* in field trials with cylindrical sticky traps. The 2-phenyl-1-ethanol (phenethanol) found in *Cucurbita maxima* blossom aroma (Andersen & Metcalf 1987) was less attractive, and attractivity was maximal for those phenyl alcohols with a C_3 side chain (Table 4.9) (Metcalf & Lampman 1989c). It was logical to evaluate the effects of incorporation of *para*-CH_3O groups into these alcohols, and 4-methoxycinnamyl alcohol was found to have low attractivity. This, we believe, is because of its extremely

Table 4.9. Attractivity of Cinnamyl Alcohol and Analogues to Adult Corn Rootworms*

Lure	1 day sticky trap catch—100 mg (mean ± S.D.)[1] *D. barberi*	*D. v. virgifera*
$C_6H_5CH{=}CHCH_2OH$ (cinnamyl alcohol)	150.7 ± 36.3c	0.2 ± 0.5a
$C_6H_5CH_2OH$ (benzyl alcohol)	11.5 ± 9.4ab	1.0 ± 1.4a
$C_6H_5CH_2CH_2OH$ (phenethanol)	28.2 ± 15.2b	1.5 ± 1.3a
$C_6H_5CH_2CH_2CH_2OH$ (phenpropanol)	125.5 ± 49.0c	1.2 ± 1.5a
$C_6H_5CH_2CH_2CH_2CH_2OH$ (phenbutanol)	8.5 ± 4.2a	1.2 ± 1.9a
control	6.0 ± 2.2a	0.2 ± 0.5a

[1] means followed by different letters are significantly different ($P \leq 0.05$).

* Data from Metcalf & Lampman (1989b).

Table 4.10. Attractivity of Methoxyphenethanol and Methoxyphenpropanol to Adult Corn Rootworms*

Lure	1 day sticky trap catch—100 mg (mean ± S.D.)[1] *D. barberi*	*D. v. virgifera*
$C_6H_5CH_2CH_2OH$	12.2 ± 5.3b	9.2 ± 9.2a
$4\text{-}CH_3OC_6H_4CH_2CH_2OH$	56.7 ± 7.5c	6.0 ± 4.2a
$C_6H_5CH_2CH_2CH_2OH$	36.2 ± 12.2c	3.2 ± 2.2a
$4\text{-}CH_3OC_6H_4CH_2CH_2CH_2OH$	15.8 ± 6.2b	3.2 ± 2.2a
control	3.5 ± 1.0a	4.6 ± 1.6a

[1] means followed by different letters are significantly different ($P \leq 0.05$).

* Data from Metcalf & Lampman (1991).

low volatility. In contrast, 4-methoxyphenpropanol and 4-methoxyphenethanol had high attractivity to *D. barberi* adults (Table 4.10), and the latter is the most effective lure yet found for this species. It is also attractive to *D. cristata*, but is unattractive to *D. virgifera*. Significant attraction of *D. barberi* in field tests was obtained with as little as 100 μg of 4-methoxyphenethanol on wicks of cylindrical sticky traps (Figure 4.7) (Metcalf & Lampman 1991). The 4-methoxyphenethanol is a natural product and has been isolated from the flowers of *Thalictrum regusum* (Ranunculaceae) and from *Aubrieta hydrida* (Cruciferae) (Metcalf & Lampman 1991).

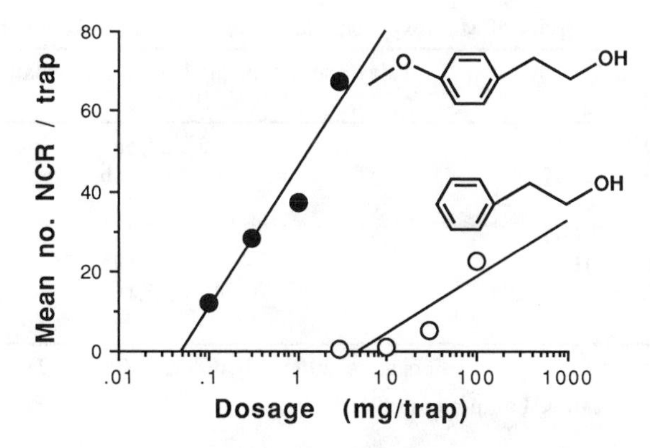

Figure 4.7. Attraction of *D. barberi* (NCR) adults to cylindrical sticky traps baited with varying doses of phenethanol and 4-methoxyphenethanol. Reprinted with permission from Metcalf & Lampman 1991.

VII. MAPPING THE ACTIVE SITES FOR KAIROMONE RECEPTORS

A. *Diabrotica barberi* Receptor

Northern corn rootworm adults are strongly attracted to phenylpropanoids with side chains containing an alcohol moiety, e.g. cinnamyl alcohol (Metcalf & Lampman 1989c). From the averages of four complete field evaluations at equivalent doses during 1988–1989, the ratio of mean numbers of *D. barberi* adults attracted as compared to cinnamyl alcohol were: phenpropanol 0.84, phenethanol 0.74, and *trans*-cinnamaldehyde 0.25. The addition of *para*-CH_3O groups substantially increased attraction so that 4-methoxyphenethanol was 3.9 times more attractive than cinnamyl alcohol and 5.3 times more attractive than phenethanol. 4-Methoxyphenpropanol was 1.13 times more attractive than phenpropanol. However, 4-methoxycinnamyl alcohol was only about 0.03 times as attractive as cinnamyl alcohol because of its very low release rate (Metcalf & Lampman 1991).

The enhanced attraction of 4-methoxyphenethanol over phenethanol, as measured by the LR values (Figure 4.7), is a function of *para*-CH_3O substitution, as attraction was greatly decreased by *ortho*-CH_3O or *meta*-CH_3O, or by *para*-substitution with F, Cl, CH_3, NO_2 or NH_2 (Table 4.11) . None of the phenethanol analogues was attractive to *D. v. virgifera* adults.

Changing the aromatic phenyl ring of cinnamyl alcohol to furylacryl

Table 4.11. Attractivity of Methoxyphenethanols to Adult Corn Rootworms*

Lure	1 day sticky trap catch—100 mg (mean ± S.D.)[1] *D. barberi*
$C_6H_5CH_2CH_2OH$	12.2 ± 5.5a
$4\text{-}CH_3OC_6H_4CH_2CH_2OH$	60.3 ± 26.8b
$3\text{-}CH_3OC_6H_4CH_2CH_2OH$	7.2 ± 3.4a
$2\text{-}CH_3OC_6H_4CH_2CH_2OH$	13.5 ± 9.3a
$C_6H_5CH{=}CHCH_2OH$	50.7 ± 50.9b
$4\text{-}CH_3OC_6H_4CH{=}CHCH_2OH$	5.2 ± 3.1a
control	4.7 ± 3.9a

[1] means followed by different letters are significantly different ($P \leq 0.05$).

* Data from Metcalf & Lampman (1991).

Table 4.12. Attractivity of Arylamines to Adult Corn Rootworms*

	1 day sticky trap catch—100 mg (mean ± S.D.)[1]	
Lure	*D. barberi*	*D. v. virgifera*
$C_6H_5CH{=}CHCH_2OH$	127.0 ± 24.2f	3.0 ± 1.4a
$C_6H_5CH_2NH_2$	57.5 ± 29.6e	1.5 ± 1.3a
$C_6H_5CH_2CH_2NH_2$	206.7 ± 32.3f	2.5 ± 1.3a
$C_6H_5CH_2CH_2CH_2NH_2$	131.5 ± 16.5f	2.2 ± 2.2a
$C_6H_5CH_2CH_2CH_2CH_2NH_2$	12.2 ± 6.1bc	1.5 ± 1.7a
$C_6H_5CH_2CH_2NHCH_3$	47.2 ± 16.3cd	1.2 ± 1.1a
$C_6H_4CH_2CH_2N^+(CH_3)_3$	9.2 ± 5.3ab	2.0 ± 1.8a
control	5.5 ± 4.7a	2.0 ± 2.4a

[1] means followed by different letters are significantly different ($P \leq 0.05$).

* Data from Metcalf & Lampman (1991).

alcohol decreased attractivity by a factor of 0.34 times, and saturation to cyclohexyl propanol reduced the factor to 0.05. Phenethylamine was 1.37 times and phenpropylamine 0.88 times as attractive as cinnamyl alcohol (Table 4.12), but the phenalkylamines were unattractive to *D. v. virgifera* or *D. u. howardi*. Phenpropylthiol ($C_6H_5CH_2CH_2CH_2SH$) was moderately attractive to *D. barberi* (Metcalf & Lampman 1989c), and we attribute the attractivity of the phenalkyl amines to the interaction of the unshared electron-pairs of the respective OH, SH, and NH_2 groups with the receptor site for the OH group of the kairomone. Cinnamyl acetate (Metcalf & Lampman 1989c) and the acetates of phenethanol and phenpropanol were significantly less active than the corresponding free alcohols (Metcalf & Lampman 1991). From this evidence we can observe the receptor site interactions of *D. barberi* with 4-methoxyphenethanol, the most effective lure, as shown in Figure 4.8.

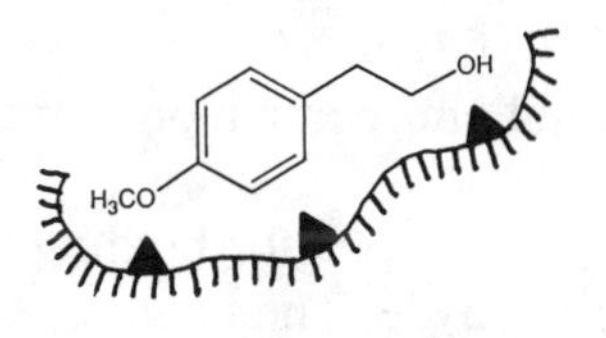

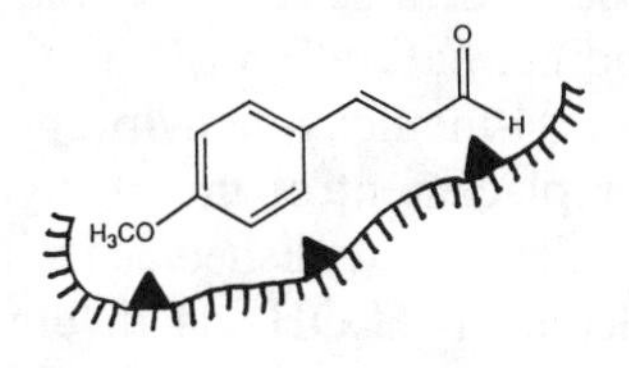

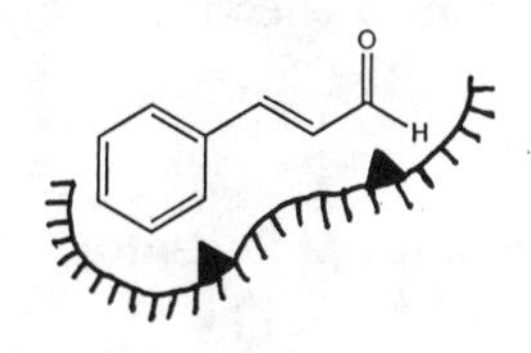

Figure 4.8. Maps of suggested receptor site interactions of *Diabrotica barberi* with 4-methoxyphenethanol, *D. v. virgifera* with 4-methoxycinnamaldehyde, and *D. u. howardi* with cinnamaldehyde.

B. *Diabrotica virgifera* Receptor

Western corn rootworm adults are strongly attracted to phenylpropanoids with side chains containing an aldehyde moiety, and 4-methoxycinnamaldehyde is the most effective attractant yet found for this species (Metcalf & Lampman 1989b). From the averages of four complete field evaluations at equivalent doses made during 1987–1988, the ratio of numbers of *D. virgifera* attracted as compared to 4-methoxycinnamaldehyde were: *trans*-cinnamyl alcohol 0.01 times and 4-methoxycinnamonitrile 0.70 times (Metcalf & Lampman 1989b). Maximal attraction was almost specifically associated with *para*-CH_3O, and the *ortho*- and *meta*-methoxycinnamaldehydes were unattractive, as were the cinnamaldehydes with *para*-substitution by F, CH_3, CH_3S, NO_2 and CF_3O (Metcalf & Lampman 1991). Cyclohexyl propanol and pentafluorocinnamaldehyde were completely unattractive.

These results indicate strong receptor binding of both the *para*-CH_3O group and the terminal aldehyde. The enhanced attractivity of 4-methoxycinnamaldehyde over cinnamaldehyde as measured by LR values is shown in Figure 4.6. The high degree of attractancy of 4-methoxycinnamonitrile (Metcalf & Lampman 1989b), which is a bioisostere of 4-methoxycinnamaldehyde, indicates the key role of C = O or C ≡ N dipoles in receptor binding. Saturation of the C = C bond of 4-methoxycinnamaldehyde, as in 4-methoxyphenylpropanol, reduced activity to about 0.01 times, and replacement of the phenyl ring as in cyclohexylpropanal or furylacrylaldehyde abolished activity. Replacement of aldehyde (C(O)H) by alcohol (CH_2OH) as in cinnamyl alcohol, phenethanol, phenpropanol, or 4-methoxyphenethanol produced compounds unattractive to *D. virgifera* (Metcalf & Lampman 1991). From this evidence, we view the receptor site interactions of *D. virgifera* with 4-methoxycinnamaldehyde, the most effective lure, as shown in Figure 4.8.

C. *Diabrotica undecimpunctata* Receptor

Trans-cinnamaldehyde is the most effective lure yet found for the Southern corn rootworm. From the averages of four replicated field evaluations during 1988–1989, the ratios of numbers of adults attracted compared to *trans*-cinnamaldehyde were: *trans*-cinnamonitrile 0.75 times, *trans*-cinnamyl alcohol 0.20 times, phenpropanol 0.55 times, 4-methoxycinnamaldehyde 0.07 times, and 4-methoxycinnamonitrile 0.05 times (Metcalf & Lampman 1989b,c). Phenethanol was slightly attractive (Lampman et al. 1987) but both 4-methoxyphenethanol and 4-methoxyphenpropanol were unattractive (Metcalf & Lampman 1991). Cinnamyl fluoride was approximately as attractive to *D. undecimpunctata* as cinnamaldehyde, and furylacrylaldehyde was 0.27 times as attractive as cinnamaldehyde. However, pentafluorocinnamaldehyde was only 0.09 times as attractive and cyclohexylpropanaldehyde was unattractive (Metcalf & Lampman 1991). Thus receptor interaction with *D. undecimpunctata* is strongly dependent upon an aromatic ring and hydrogen bonding, and upon a terminal aldehyde (C = O) or nitrile (C ≡ N) group, and moderately dependent upon side chain unsaturation. From this evidence, we view the receptor site interactions of *D. undecimpunctata* with cinnamaldehyde, the most attractive kairomone lure as shown in Figure 4.8.

Table 4.13. Persistence of Lure Doses on Cylindrical Sticky Traps for Adult *Diabrotica barberi* (NCR) and *D. v. virgifera* (WCR)*

Days after Dosing	Mean 24 hr trap catch–100 mg[1]				
	Control	Eugenol	Cinnamyl alcohol	4-methoxy-cinnamaldehyde	TIC mixture
1 NCR	6.5a	45.5b	151.7c	5.5a	48.5b
WCR	2.0a	1.5a	3.7a	124.5b	130.5b
3 NCR	1.7a	26.0b	28.2b	2.0a	18.5b
WCR	1.5a	1.0a	2.2a	151.7c	68.2b
5 NCR	0.2a	13.2b	12.0b	5.0a	18.0b
WCR	2.2a	4.5a	5.0a	82.5b	53.5b
7 NCR	1.0a	20.0b	3.7a	1.7a	0.5a
WCR	1.0a	0.7a	0.7a	52.6b	22.9b
10 NCR	2.0a	0.7a	1.1a	1.0a	1.8a
WCR	2.5a	2.1a	0.7a	44.7b	2.0a

[1] means followed by different letters are significantly different ($P \leq 0.05$).

* Data from Lampman et al. (1991).

VIII. KAIROMONES FOR MONITORING AND CONTROLLING DIABROTICITES

Kairomone attractants for the several species of adult corn rootworms and cucumber beetles are powerful tools for insect pest management. Adult rootworm beetle lures used with cylindrical sticky traps show a linear relationship of trap catch vs. log of lure dosage (Andersen & Metcalf 1986, Metcalf & Lampman 1989b, 1991, Lewis et al. 1990). This extends over a range of 10 to 500 beetles per trap, and from 0.03 to 100 mg of lure dose per trap (Figures 4.6, 4.7). Lures such as 4-methoxycinnamaldehyde for *D. v. virgifera*, 4-methoxyphenethanol for *D. barberi* and the generally attractive TIC mixture (equal parts of 1,2,4-trimethoxybenzene, indole, and cinnamaldehyde) used at 100 mg per sticky trap are effective in luring the corn rootworm adults upwind from distances as great as 100 m (Lampman et al. 1991). Studies of the longevity of effectiveness of cylindrical sticky traps baited with 100 mg of these lures show that they can be used to trap adult corn rootworms for up to two weeks (Table 4.13) under exposure to hot mid-western corn growing conditions (Lampman et al. 1991). The different field longevities of the several lures are the results of different release rates and behavioral thresholds for rootworm beetle response.

A. Efficiency of Corn Rootworm Lures in Adult Trapping

Cylindrical sticky traps baited with efficient lures, such as TIC, 4-methoxyphenethanol, 4-methoxycinnamaldehyde, cinnamaldehyde, and cinnamyl alcohol, are effective and sensitive monitoring tools for the quantitative monitoring of adult rootworm beetle populations in corn (Metcalf & Lampman 1989b, c, 1991, Lampman et al.1991). Correlation of sticky trap catches with rootworm populations counted *in situ* on corn plants has often shown significant trap catches when no rootworm beetles could be detected by sampling 100 to 500 plants. Extensive field trapping in a 120 acre field of hybrid seed corn heavily infested with both *D. v. virgifera* and *D. barberi*, over the period of Aug. 8–22, 1990, afforded an opportunity to compare the efficiency of plant counting with kairomone trap counting using TIC lure at 100 mg per yellow "Solo Cup" sticky trap (Levine & Metcalf 1988). Two thirds of the field was treated with cucurbitacin-volatile kairomone-carbaryl bait, so that after an initial pretreatment monitoring, it was possible to compare the results of plant counts (100 plants counted in each corner of control and treatment areas) with sticky trap counts. The results from 22 complete observations are shown in Figure 4.9. The correlation coefficient $R^2 = 0.590$ ($P \leqq 0.001$) showed the reliability of the correlation between actual plant counts and sticky trap counts. Thus sticky trap counts with TIC attractant are approximately 1000 times more effective in measuring adult rootworm beetle densities in corn than actual plant counts. The trap becomes less efficient at very high beetle densities because all of the sticky area becomes impacted with beetles. By dividing the mean number of beetles per trap by the whole plant count, this and other, similar evaluations have shown that over a 24 hour period, each TIC baited sticky trap attracted WCR beetles from 500–1000 plants. In similar experiments, 4-methoxyphenethanol at 100 mg per trap sampled NCR beetles from the equivalent of about 4000 corn plants (Lampman et al. 1991).

B. Cucurbitacin-containing Baits

The tetracyclic cucurbitacins B and E are of high molecular weight, and their lack of appreciable volatility makes them ineffective as attractants. However, the extraordinary arrestant and phagostimulant properties of these kairomones suggested their use not only as tools for studying Diabroticite evolution and behavior, but also in integrated pest management of cucumber beetles and corn rootworms: (a) for monitoring rootworm beetle populations, (b) as components of poison baits, and (c) as trap crops (Rhodes et al. 1980).

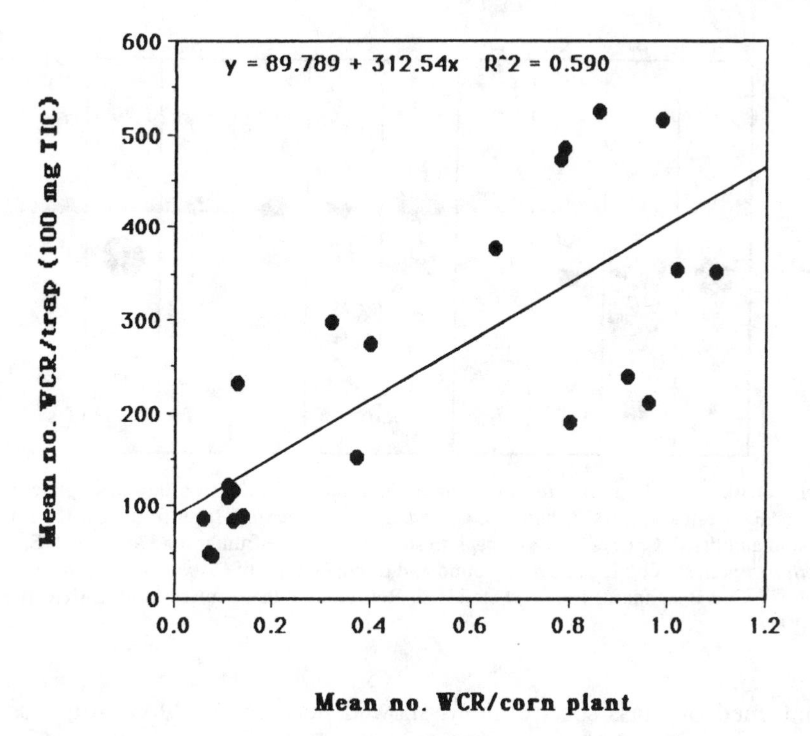

Figure 4.9. Correlation of plant count of *D. v. virgifera* adults (WCR) with adults captured by cylindrical sticky traps baited with TIC attractant (trimethoxybenzene, indole, cinnamaldehyde).

Wild Cucurbitaceae do not provide dependable sources of cucurbitacins because they are difficult to grow and fruiting is often dependent upon photoperiod. Therefore, Cuc-containing genes were transferred to domesticated cultivars (Rhodes et al. 1980). After experimenting with 18 *Cucurbita* species and > 30 hybrids (Metcalf & Rhodes 1990), two promising hybrids were developed: *Cucurbita texana* × *C. pepo* cv. Zucchini (TEX × PEP) and *C. andreana* × *C. maxima* cv. Macre (AND × MAX). The TEX × PEP hybrid produced semibush plants with leaves, fruit, and blossoms resembling *C. pepo* cv. Zucchini. The average weight of the fruits was 0.73 kg. The AND × MAX hybrid produced long-vined plants with leaves, fruits, and blossoms resembling *C. maxima.* The average weight of the fruits was 3.9 kg (Rhodes et al. 1980).

The "beetle print" feeding assay (Figure 4.10) showed that the fruits of the two hybrids contained relatively large amounts of Cuc phagostimulants, the AND × MAX hybrid Cucs B-D, and the TEX × PEP hybrid Cucs E-I and glycoside. Quantitative assays by ultraviolet spectrometry,

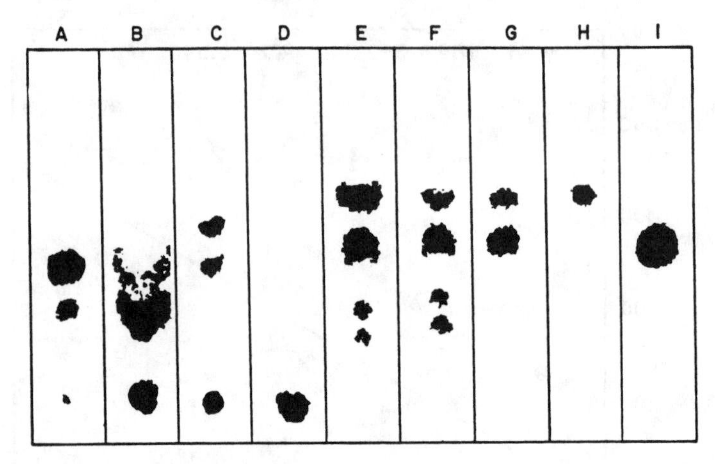

Figure 4.10. "Beetle prints" from feeding of *D. u. howardi* adults on thin-layer plates of extracts of fresh and dry Cuc baits: A, *C. andreana* × *C. maxima* fresh fruit; B, same fruit ground and dried; C, *C. texana* × *C. pepo* fresh fruit; D. same fruit ground and dried; E, *C. foetidissima* fresh root; F, same root ground and dried; G, juice of *C. andreana* × *C. maxima* fruit; H, Cuc B standard; I. Cuc D standard. Reprinted with permission from Metcalf et al. 1987.

confirmed by mass spectrometry, showed that AND × MAX fruit contained a total of 0.126% Cucs and TEX × PEP a total of 0.051% Cucs (Table 4.14) (Rhodes et al. 1980).

Cut bitter *Cucurbita* fruits have been recognized as highly effective arrestants for Diabroticites since Contardi (1939) showed that *D. speciosa* preferred to feed on the bitter squash *C. andreana* and ignored its close relative, the sweet cultivar *C. maxima.* When equal numbers of split fruits were assayed, 99% of the beetles were found feeding on the bitter squash. Sharma and Hall (1973) conducted similar experiments and found *D. u. howardi* beetles were attracted to cut bitter *C. foetidissima* fruits about 15 times more abundantly than to the fruits of the sweet cultivar *C. pepo.* Howe et al. (1976) showed that *D. v. virgifera* adults had the following mean preferences for cut *Cucurbita* fruits: *C. maxima* 0.3, *C. pepo* 0.5, *C. andreana* 13.3 and *C. texana* 14.0. Cut fruits and homogenates of the TEX × PEP and AND × MAX hybrids were very effective in arresting both *D. u. howardi* and *D. v. virgifera* adults, which simply ate all the fruit and flew away. However, when 0.1 g of methomyl insecticide was dusted on the cut halves at approximately 0.03–0.1% of the fruit weight, adults of both *D. u. howardi* and *D. v. virgifera* fed avidly and were killed within 2–5 minutes. The average numbers of dead rootworm beetles (n = 10) found after 48 hours exposure in a cucurbit plot was *C. andreana* 325 ± 92, *C. andreana* × *C. maxima* hybrid 342 ± 179, *C. texana* 259

Table 4.14. Cucurbitacin Content of Fresh Fruits and Dried Baits from *C. andreana* × *C. maxima* and *C. texana* × *C. pepo* Hybrid Fruits and *C. foetidissima* Root.*

Source	Cucurbitacins (mg per g)						
	B	D	E	I	Unknown	Glycoside	Total
AND × MAX fruit							
F_1 fresh	1.17	0.09					1.26
F_1 dried	0.90	0.64				2.92	4.46
F_2 dried	0.24					1.84	2.08
TEX × PEP fruit							
F_1 fresh			0.23	0.09		0.19	0.51
F_1 dried			1.44	1.36		3.15	5.95
F_2 dried			1.36			1.63	2.99
C. foetidissima root							
fresh			0.28	1.72	0.51	0.59	3.10
dried			1.76	2.07			3.83

* Reprinted with permission from Metcalf et al. (1987).

± 100, and *C. texana* × *C. pepo* hybrid 330 ± 86 (Rhodes et al. 1980). The rapid toxic action of the baits prevented the beetles from eating appreciable quantities so that the toxic fruits remained effective for at least three weeks, and a single methomyl-treated fruit half was estimated to have killed in excess of 2000 rootworm beetles (Rhodes et al. 1980, Metcalf 1985). Such fruit baits are effective monitoring tools for Diabroticite beetles.

1. Dried Cucurbita *Fruit Baits*

The development of high yielding *Cucurbita* hybrids with fruits containing substantial amounts of Cucs (Table 4.14) made it possible to experiment with dried, ground bitter *Cucurbita* baits for the control and management of rootworm adults in cucurbits and corn. Dried *Cucurbita* F_1 and F_2 baits were produced from the fruits of the hybrids of AND × MAX and TEX × PEP (Metcalf et al. 1987). The fruits were split, partially air dried, then thoroughly dried in a forced-air oven at 70° C and ground in a Wiley Mill. The AND × MAX bait contained 8.9% solids and 34% passed a 2 mm screen and the TEX × PEP baits contained 9.1% solids and 70% passed a 2 mm screen. These baits were assayed for Cuc content by the "beetle print" assay (Figure 4.10) and by ultraviolet spectrometry (Table 4.14). For the hybrids, total Cuc content of F_2 fruit decreased because of segregation of the dominant Bi gene controlling Cuc synthesis, e.g. TEX × PEP F_1 fruit bait contained ca. 0.6% and the F_2 bait 0.3% Cucs; and the AND × MAX F_1 fruit bait contained ca. 0.45% Cucs and the F_2 bait 0.21%

(Table 4.14). Drying of bitter *Cucurbita* fruits substantially increased the percentage of Cucs, and also altered their composition. Fresh AND × MAX fruits contained a preponderance of Cuc B, but the dried bait contained much larger amounts of Cuc D (deacetoxy Cuc B) and Cuc-glycosides. Fresh TEX × PEP fruit contained a mixture of Cuc E, I and glycosides but the dried bait contained a much higher amount of glycosides (Table 4.14). Nevertheless *D. balteata*, *D. v. virgifera*, and *D. u. howardi* adults fed avidly on the Cuc mixtures extracted from the dried baits as shown by the "beetle print" assay of Figure 4.10. The drying and grinding process evidently resulted in the liberation of enzymes catalyzing these changes by cellular disruption (Metcalf et al. 1987).

2. Formulated Cucurbitacin Baits

The most efficient use of Cuc arrestants and phagostimulants in toxic bait may be through formulation on an inert pelleted carrier. This was investigated by impregnating 10-mesh (71% passed 2 mm-screen) corn grits (Pesticide Carrier Grits 980, Illinois Cereal Mills) with 0.1% methomyl insecticide, and then further impregnating the carrier with a chloroform extract of AND × MAX fruit containing 0.12% Cucs B-D to give a range of Cuc concentrations. Petri dish halves containing 300 mg of the bait (equivalent to 17 kg per ha) were exposed to *D. u. howardi* adults in a cucurbit plot for one day. The mean (± SD) numbers of dead beetles (n = 10) were: control, no Cucs - 0; 0.0036% Cucs - 7.9 ± 11.1a; 0.012% Cucs - 25.9 ± 18.5ab; 0.036% Cucs - 37.9 ± 23.3b; and 0.12% Cucs - 65.9 ± 28.4c (Metcalf et al. 1987). The minimum effective Cuc concentration was about 0.012% or the equivalent of about 2 g of Cucs per ha.

C. Field Experiments with Cucurbitacin Baits

A variety of insecticides have been evaluated as components of the Cuc-containing baits. In small plot evaluations with TEX × PEP F_1 bait applied to the ground in sweet corn at 1 g per m^2, the carbamates carbofuran, bendiocarb, methomyl and carbaryl at 0.1% were the most effective in decreasing order. Isofenphos at 0.1% was less effective, and the pyrethroids permethrin, cypermethrin, fenvalerate, and flucythrinate at 0.01% were ineffective, apparently because of repellency (Metcalf et al. 1987). In 0.04 ha plots of sweet corn, where TEX × PEP F_2 bait was broadcast over the plants at 33 kg per ha, 0.1 methomyl bait produced 67.6% mortality of *D. v. virgifera* and 93.7% reduction in precount populations. The

Table 4.15. Effectiveness of *Cucurbita texana* × *C. pepo* Dried Bait with 0.1 % Methomyl, Broadcast over Sweet Corn*

Application rate		No. corn rootworm beetles per 0.01 ha plot			
Bait (kg/ha)	Methomyl (g/ha)	Precount		Post count (% mortality)—1 day	
		D. v. virgifera	*D. u. howardi*	*D. v. virgifera*	*D. u. howardi*
1.1	11.2	46	125	9 (20)	26 (21)
3.3	33.6	308	281	123 (40)	229 (81)
11	112	256	301	189 (77)	350 (100)
11	11.2	133	213	54 (41)	113 (53)
33	33.6	84	321	82 (98)	1,322 (100)
untreated	—	233	57	4 (2)	3 (5)

* Data from Metcalf et al. (1987).

comparable values for 0.1% isofenphos bait were 99.9% mortality and 77.5% reduction.

Application rates of Cuc Baits. TEX × PEP F_1 bait with 0.1 or 1.0% methomyl was broadcast over 0.01 ha sweet corn plots at rates of 1.1 to 33 kg per ha. The plots were heavily infested with *D. v. virgifera* and *D. u. howardi* adults. The effectiveness of the bait applications was based on pre- and post-treatment counts of the beetles per plot, and on numbers of dead beetles on the ground 1 day after application. As shown in Table 4.15, 0.1% methomyl bait applied at 33 kg per ha was optimally effective, producing 90–100% mortality of both species of rootworm adults at an application rate of 33.6 g of methomyl per ha (13.6 g per acre). This rate of insecticide application in the Cuc baits represents a reduction of 97% over the conventional dosage of 1.1 kg per ha (1 lb per acre) for aircraft spraying to control adult corn rootworms. Cuc bait containing 1% methomyl applied at 3.3 kg per ha, to give an equivalent dosage of 33.6 g of methomyl per ha, was less effective (Table 4.15) because of less efficient distribution of the bait (Metcalf et al. 1987).

Cuc baits with volatile kairomone attractants. The effectiveness of the Cuc baits is dependent upon random search by Diabroticite beetles, and their arrest and phagostimulation to feed on the insecticide by direct relationship between Cuc content and bait effectiveness (Metcalf et al. 1987). The present availability of an array of volatile kairomone attractants for Diabroticites affords opportunities to fortify the Cuc containing bait with long range attractants that should increase bait efficiency. This factor was evaluated in a preliminary way by adding increasing amounts of TIC attractant mixture (1,2,4-trimethoxybenzene-indole-*trans*-cinnamaldehyde) to the corn cob grits bait containing 0.3% carbaryl and 5% *C. foetidissima* root powder. Cylindrical sticky traps were treated with 1 g of this bait containing from 1 to 5 mg of TIC and the numbers of *D.*

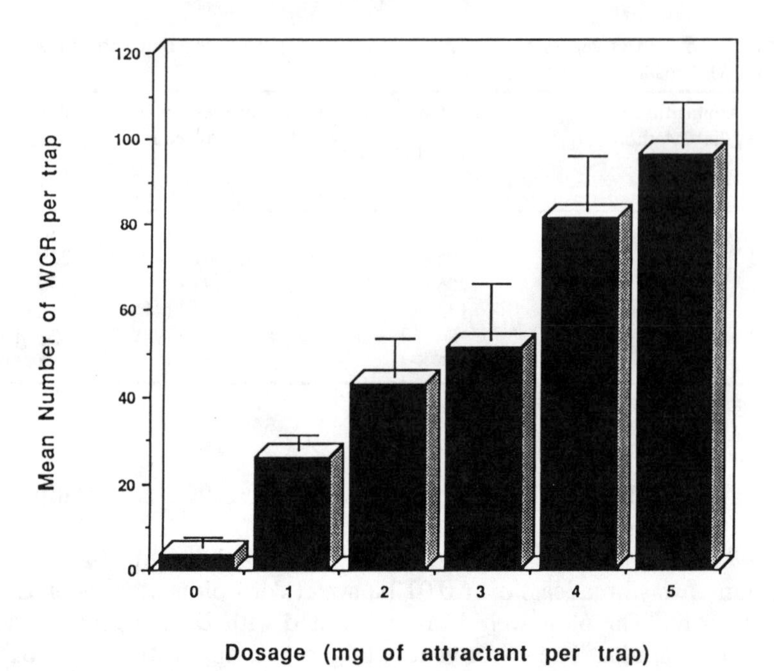

Figure 4.11. Attraction of *D. v. virgifera* adults (WCR) to cylindrical sticky traps baited with granular Cuc-containing bait and varying amounts of TIC attractant (trimethoxybenzene, indole, cinnamaldehyde). (From Lampman et al. 1991).

v. virgifera adults caught in the traps were determined in replicated tests (n = 4) at the border of a field of hybrid corn (Lampman et al. 1991). As shown in Figure 4.11, the numbers of rootworm adults attracted to the bait were increased about 10 times by a TIC concentration of 0.1% and to 50 times by a TIC concentration of 0.5% in the Cuc bait.

This account of the usefulness of kairomones for the control of Diabroticite beetles is approximately "state of the art". Cucurbitacin-containing baits have been shown to produce as much as 98–100% kill of adult Diabroticites infesting cucurbits and corn at application rates as low as 10 kg per ha and insecticide doses of 13 g per ha.

Aerial application of a commercial corn cob grits bait containing 0.01% Cucs, 0.3% carbaryl and 0.1% each of the volatile attractants trimethoxybenzene, indole, cinnamaldehyde, cinnamyl alcohol, phenpropanol, 4-methoxycinnamaldehyde, and 4-methoxyphenethanol (Nemesis, Biocontrol Ltd) at approximately 10 kg per ha to an 80 acre cornfield resulted in 98 to 100% mortality of *D. barberi* and *D. v. virgifera* adults within 24 hours. The incorporation of the volatile kairomones into this Cuc bait increased the kill of rootworm adults by at least 2.5 times (Lampman et al. 1991).

Much remains to be learned about timing of these applications in corn to give maximum kill of gravid female *D. barberi* and *D. v. virgifera*. Formulations of the Cuc baits can be improved and the optimal Cuc content, insecticide content, and rate of application need further study.

The volatile lure additives for Cuc baits can be improved. Based on attractant release rates and persistence, a mixture of 4-methoxycinnamaldehyde and 4-methoxyphenethanol might be a superior bait lure for *D. barberi* and *D. v. virgifera* (Lampman et al. 1991).

REFERENCES

Andersen, J.F. 1987. Composition of the floral odor of *Cucurbita maxima* (Duchesne (Cucurbitaceae). J. Agric. Food Chem. 35: 60–62.

Andersen, J.F. and R.L. Metcalf 1986. Identification of a volatile attractant for *Diabrotica* and *Acalymma* species from the blossoms of *Cucurbita maxima* Duchesne. J. Chem. Ecol. 12: 687–699.

Andersen, J.F. and R.L. Metcalf 1987. Factors influencing the distribution of *Diabrotica* spp. in the blossoms of cultivated *Cucurbita* spp. J. Chem. Ecol. 14: 681–689.

Andersen, J.F., R.D. Platner and D. Weisleder. 1988. Metabolic transformations of cucurbitacins by *Diabrotica virgifera* LeConte and *D. undecimpunctata howardi* Barber. Insect Biochem. 18: 71–77.

Chamblis, O.L. and C.M. Jones. 1966. Cucurbitacins: specific insect attractants in Cucurbitaceae. Science 153: 1392–1393.

Contardi, H. 1939. Estudios geneticos en "Cucurbita" y consideraciones agronomicas. Physis 18: 331–347.

Crowson, R.A. 1981. "The Biology of the Coleoptera." Academic Press N.Y.

Curtis, P.S. and P.M. Meade. 1971. Cucurbitacins from Cruciferae. Phytochemistry 10: 3081–3083.

DaCosta, C.P. and C.M. Jones. 1971a. Resistance in cucumber, *Cucumis sativa* to three species of cucumber beetles. Hort. Science 6: 340–342.

DaCosta, C.P. and C.M. Jones. 1971b. Cucumber beetle resistance and mite susceptibility controlled by the bitter gene in *Cucumis sativa* L. Science 172: 1145–1146.

David, A. and D.K. Vallance. 1955. Bitter principles of cucurbitaceae. J. Pharm. Pharmac. 7: 295–296.

Dryer, D.L. and E.K. Trousdale. 1978. Cucurbitacins in *Purshia tridentata.* Phytochemistry 17: 325–326.

Enslin, P.R., F.J. Joubert and S. Rehm. 1956. Bitter principles of the Cucurbitaceae III. Elaterase an active enzyme for the hydrolysis of bitter principle glycosides. J. Sci. Food Agric. 7: 646–655.

Ferguson, J.E., D.C. Fischer and R.L. Metcalf. 1983. A report of cucurbitacin poisoning in humans. Cucurbit. Genet. Coop. Rep. 6: 73–74.

Ferguson, J.E., E.R. Metcalf, R.L. Metcalf and A.M. Rhodes. 1983. Influence of cucurbitacin content in cotyledons of Cucurbitaceae cultivars upon feeding behavior of Diabroticina beetles (Coleoptera: Chrysomelidae). J. Econ. Entomol. 76: 47–51.

Ferguson, J.E. and R.L. Metcalf. 1985. Cucurbitacins:plant derived defense compounds for Diabroticina (Coleoptera: Chrysomelidae. J. Chem. Ecol. 11: 311–318.

Ferguson, J.E., R.L. Metcalf and D.C. Fischer. 1985. Disposition and fate of cucurbitacin B in five species of Diabroticina. J. Chem. Ecol. 11: 1307–1321.

Fischer, J.R., T.F. Branson and G. R. Sutter. 1984. Use of common squash cultivars, *Cucurbita* spp. for mass collection of corn rootworm beetles, *Diabrotica* spp. (Coleoptera: Chrysomelidae). J. Kansas Entomol. Soc. 57: 409–412.

Friedrich, H. 1976. Phenylpropanoid constituents of essential oils. Lloydia 39: 1–7.

Fronk, W.D. and J.H. Slater. 1956. Insect fauna of cucurbit flowers. J. Kansas Entomol. Soc. 29: 141–145

Guha, J. and S.P. Sen. 1975. The cucurbitacins - a review. Ind. Biochem. J. 12:28.

Howe, W.L., J.R. Sanborn and A.M. Rhodes. 1976. Western corn rootworm adults and spotted cucumber beetle associations with Cucurbita and cucurbitacins. Environ. Entomol. 5: 1043–1048.

Kevan, P.G. and H.G. Baker. 1983. Insects as flower visitors and pollinators. Annu. Rev. Entomol. 28: 407–453.

Krysan, J.L., T.F. Branson, R.F.W. Schroder and W.E. Steiner, Jr. 1984. Elevation of *Diabrotica sicuanica* (Coleoptera: Chrysomelidae) to the species level with notes on the altitudinal distribution of *Diabrotica* species in the Cuzco department of Peru. Entomol. News 93: 91–98.

Krysan, J.L., I.C. McDonald and J.H. Tumlinson. 1989. Phenogram based on allozymes and its relationship to classical biosystematics and pheromone structure among eleven Diabroticites (Coleoptera; Chrysomelidae). Ann. Entomol. Soc. Amer. 82: 574–581.

Krysan, J.L. and R.F. Smith. 1987. Systematics of the *virgifera* species group of *Diabrotica* (Coleoptera: Chrysomelidae: Galerucinae). Entomography 5: 375–484.

Ladd, Jr, T.L. 1984. Eugenol-related attractants for the northern corn rootworm (Coleoptera: Chrysomelidae). J. Econ. Entomol. 77: 339–341.

Ladd, T.L., B.R. Stinner and H.R. Krueger. 1983. Eugenol, a new attractant for the northern corn rootworm (Coleoptera: Chrysomelidae). J. Econ. Entomol. 76: 1049–1051.

Lampman, R.L. and R.L. Metcalf. 1987. Multicomponent kairomone lures for southern and western corn rootworms (Coleoptera: Chrysomelidae: *Diabrotica* spp.). J. Econ. Entomol. 80: 1137–1142.

Lampman, R.L. and R.L. Metcalf. 1988. The comparative response of *Diabrotica* species (Coleoptera: Chrysomelidae) to volatile attractants. Environ. Entomol. 17: 644–648.

Lampman, R.L., R.L. Metcalf and J.F. Andersen. 1987. Semiochemical attractants of *Diabrotica undecimpunctata howardi* Barber, southern corn rootworm, and *Diabrotica virgifera virgifera* LeConte, the western corn rootworm (Coleoptera: Chrysomelidae). J. Chem. Ecol. 13: 959–975.

Lampman, R.L., R.L. Metcalf, Lesley Deem-Dickson and C.D Reid. 1991. Attraction of Diabroticites (Coleoptera: Chrysomelidae) to a multicomponent lure and implications for corn rootworm baits. Environ. Entomol. In Press.

Lavie, D. and E. Glotter. 1971. The cucurbitacins, a group of tetracyclic triterpenes. Forts. Chem. Organ. Naturstoffe. 29: 306–362.

Levine, E. and R.L. Metcalf. 1988. Sticky attractant traps for monitoring corn rootworm beetles. Ill. Nat. Hist. Survey Rept. No. 279, Sept.

Lewis, P.A., R.L. Lampman and R.L. Metcalf. 1990. Kairomonal attractants for *Acalymma vittatum* (Coleoptera: Chrysomelidae). Environ. Entomol. 19: 9–14.

Maulik, S. 1936. "Coleoptera, Chrysomelidae (Galercucinae). Fauna of British India." Taylor and Francis, London.

Metcalf, C.L., W.P. Flint and R.L. Metcalf. 1962. "Destructive and Useful Insects.", 4th ed. McGraw-Hill, N.Y.

Metcalf, R.L. 1979. Plants, chemicals, and insects. Some aspects of coevolution. Bull. Entomol. Soc. Amer. 25 (1): 30–35.

Metcalf, R.L. 1985. Plant kairomones and insect pest control. Bull. Ill. Nat. Hist. Surv. 33: 175–198.

Metcalf, R.L. 1986. Coevolutionary adaptations of rootworm beetles (Coleoptera: Chrysomelidae) to cucurbitacins. J. Chem. Ecol. 12: 1109–1124.

Metcalf, R.L., J.E. Ferguson, R.L. Lampman and J.F. Andersen. 1987. Dry cucurbitacin-containing baits for controlling Diabroticite beetles (Coleoptera; Chrysomelidae). J. Econ. Entomol. 80: 870–875.

Metcalf, R.L. and R.L. Lampman. 1989a. Chemical ecology of Diabroticites and Cucurbitaceae. Experientia 45: 240–247.

Metcalf, R.L. and R.L. Lampman. 1989b. Estragole analogues as attractants of *Diabrotica* species (Coleoptera: Chrysomelidae) corn rootworms. J. Econ. Entomol. 82: 123–129.

Metcalf, R.L. and R.L. Lampman. 1989c. Cinnamyl alcohol and analogues as attractants for the adult northern corn rootworm *Diabrotica barberi* (Coleoptera: Chrysomelidae). J. Econ. Entomol. 82: 1620–1625.

Metcalf, R.L. and R.L. Lampman. 1991. Evolution of Diabroticite (Coleoptera: Chrysomelidae) receptors for *Cucurbita* blossom volatiles. Proc. Natl. Acad. Sci. (USA) 88:

Metcalf, R.L., R.A. Metcalf and A.M. Rhodes. 1980. Cucurbitacins as kairomones for diabroticite beetles. Proc. Natl. Acad. Sci. (USA) 77: 3769–3772.

Metcalf, R.L. and A.M. Rhodes. 1990. Pp. 167–182 in D.M. Bates R.W. Robinson and C. Jeffrey, eds. "Biology and Utilization of the Cucurbitaceae." Cornell U. Press, Ithaca, N.Y.

Metcalf, R.L., A.M. Rhodes, R.A. Metcalf, J.E. Ferguson, E. R. Metcalf, and P-y Lu. (1982). Cucurbitacin contents and Diabroticites (Coleoptera: Chrysomelidae) feeding upon *Cucurbita* spp. Environ. Entomol. 11: 931–937.

Morgan, A.C. and S.E. Crumb. 1928. Notes on chemotropic responses of certain insects. J. Econ. Entomol. 21: 913–920.

Nielson, J.K., M. Larsen and H.J. Sorenson. 1977. Cucurbitacin E and I in *Iberis amara* feeding inhibitors for *Phyllotreta nemorum.* Phytochemistry 16: 1519–1522.

Nishida, R. and H. Fukami. 1990. Sequestration of distasteful compounds by some pharmacophagous insects. J. Chem. Ecol. 16: 151–164.

Nishida, R., H. Fukami, Y. Tanaka, P. Magalhaes, M. Yokoyama, and A. Blumenschein. 1986. Isolation of feeding stimulants of Brazilian leaf beetles (*Diabrotica speciosa* and *Ceratoma arcuata*) from the roots of *Ceratosanthes hilariana.* Agric. Biol. Chem. (Japan) 50: 2831–2836.

Pal, A.B., K. Srinivasan, G. Bharatan and M. U. Chadradana. 1978. Location of sources of resistance to the red pumpkin beetle *Rapidapalpa foveicollis* Lucas among pumpkin germ plamsa. J. Entomol. Res. 2: 148–153.

Price, P. 1984. "Insect Ecology", 2nd. ed., John Wiley & Sons, N.Y.

Rehm, S. 1960. Die Bitterstoffe der Cucurbitaceae. Ergeb. Biol. 22: 106–136.

Rehm, S., P.A. Enslin, A.D.J. Meeuse and J.H. Wessels. 1957. Bitter principles of the Cucurbitaceae VII. The distribution of bitter principles in the plant family. J. Sci. Food Agric. 8: 679–686.

Rhodes, A.M., R.L. Metcalf and E.R. Metcalf 1980. Diabroticite response to cucurbitacin kairomones. J. Amer. Soc. Hort. Sci. 105: 838–842.

Rhymal, K.S., O.L. Chambliss, M.D. Bond and D.A. Smith. 1984. Squash containing toxic cucurbitacin compounds occurring in California and Alabama. J. Food Protection 47: 270–271

Robinson, R.W., H.M. Munger, T.W. Whitaker and G.W. Bohn. 1976. Genes of the Cucurbitaceae. Hort. Sci. 11: 554–568.

Schabort, J.C. and H.L. Teijema. 1968. The role of cucurbitacin Δ^{23} reductase in the breakdown of toxic bitter principles in *Cucurbita maxima.* Phytochemistry. 7: 2107–2110.

Schwartz, H.M., S.I. Biedron, M.N. von Holdt and S. Rehm. 1964. A study of some plant esterases. Phytochemistry. 3: 189–200.

Sharma, G.D. and C.V. Hall. 1973. Relative attractance of spotted cucumber beetles to fruits of fifteen species of Cucurbitaceae. Environ. Entomol. 2: 154–156.

Sinha, A.K. and S.S. Krishna. 1970. Further studies on the feeding behavior of *Aulacophora foveicollis* on cucurbitacin. J. Econ. Entomol. 63: 333–334.

Smith, R.F. and J.F. Lawrence. 1967. "Clarification of the Status of Type Specimens of Diabroticites (Coleoptera: Chrysomelidae: Galerucini)." Univ. Calif. Press, Berkeley, CA.

Stroesand, G.S., A. Jaworski, S. Shannon and R.W. Robinson. 1985. Toxicologic response in mice fed *Cucurbita* fruit. J. Food Protection 48: 50–51.

Takizawa, H. 1978. Notes on Taiwanese Chrysomelidae I. Kontyu (Tokyo) 46: 123–134.

Watt, J.M. and M.G. Breyer-Brandwyk. 1962. "The Medicinal and poisonous plants of Southern and Eastern Africa.", 2nd. ed. E. and S. Livingston, Edinburgh, U.K.

Wierman, R. 1970. Die Synthese von Phenylpropanen wahrend der Pollentwicklung. Planta 95: 133–145.

Wierman, R. 1981. Secondary plant products of plants, pp. 85–116 in P.A. Stumpf and E.E. Conn. eds. "Biochemistry of Plants.", vol. 7. Academic Press, N.Y.

Wilcox, J.A. 1972. "Chrysomelidae: Galerucinae: Luperini. Coleopterum Catalogus Supplementa.", Pars 78, Fasc. 3, 2nd. ed. Junk s'Gravenhagen, Netherlands.

Yaro, N.D., J.D. Krysan and T.L. Ladd, ur. 1987. *Diabrotica cristata* (Coleoptera: Chrysomelidae): attraction to eugenol and related compounds campared with *D. barberi* and *D. virgifera virgifera*. Environ. Entomol. 16: 126–128.

5

FRUIT FLIES OF THE FAMILY TEPHRITIDAE

I. INTRODUCTION

The Tephritidae (Diptera), or true fruit flies, are medium sized species with spotted or banded wings whose larvae typically are legless, headless maggots that tunnel in the interior of fruits or vegetables, while the adults frequent vegetation and blossoms. There are more than 1000 described species, including 700 *Dacus* (Dacinae), 200 *Anastrepha* (Trypetinae), 50 *Rhagoletis* and 100 *Ceratitis* (Ceratitinae). The great majority of these are monophagous or stenophagous specialists breeding in the fruiting bodies of many families of tropical and subtropical plants. Some of these specialists, such as the apple maggot *R. pomonella*, the European cherry fruit fly *R. cerasi*, the walnut husk fly *R. completa*, the Mexican fruit fly *Anestrepha ludens*, the olive fruit fly *D. oleae*, and the banana fruit fly *D. musae*, are important pests because they attack specific cultivars of high value. A few species of Tephritidae are polyphagous generalists and are major crop pests wherever they have been introduced in the tropics and subtropics. These include the Mediterranean fruit fly *C. capitata* with 253 host plant species, the oriental fruit fly *D. dorsalis* with 173 hosts, the melon fly *D. cucurbitae* with at least 125 hosts, and the Queensland fruit fly *D. tryoni* with 106 hosts. These major pest species are frequent importations into suitable climatic areas such as California, Florida, and Texas, which result in widespread plant quarantine measures. Hundred of millions of dollars have been expended in eradication campaigns in California and Florida. The proliferation of global air travel, where no place is more than a day away, makes infestation and reinfestation virtual certainties.

There are as many as 100 species of fruit flies of the genera *Anastrepha*, *Ceratitis*, *Dacus* and *Rhagoletis* that are of potential or real importance as threats to agriculture in the United States (Carey & Dowell 1989). A summary of the most important pest species of *Dacus* and their hosts is given in Table 5.1.

Table 5.1. *Dacus* Fruit Flies That are Pests of Agriculture*

Dacus spp.	Common name	Hosts	Location	Male lure[1]
(*Bactrocera*) *aquilonis* (May)		citrus, guava, peach	Australia	CL
(*Paratridacus*) *atrisetosus* (Perkins)		cucurbits	New Guinea	Un
(*Dacus*) *bivitattus* (Bigot)	pumpkin fly	cucurbits, cherry, coffee	Nigeria, Zimbabwe	CL
(*Bactrocera*) *breviaculeus* Hardy		guava	Australia	CL
(*Bactrocera*) *bryoniae* (Tryon)		banana, mango, passion fruit, capsicum	Australia, New Guinea	CL
(*Bactrocera*) *cacuminatus* (Hering)	solanum fly	capsicum, tomato	Australia	ME
(*Zeugodacus*) *caudatus* F.		chili, guava, mango, papaya	Malaysia, Thailand	CL
(*Didacus*) *ciliatus* Loew	Ethiopian fruit fly	cucurbits	Nigeria, Ethiopia, Egypt, India, Pakistan	Un
(*Afrodacus*) *correctus* (Bezzi)	guava fruit fly	citrus, mango, peach, guava, zapota	India, Pakistan, Thailand	ME
(*Austrodacus*) *cucumis* French	cucumber fly	cucurbits, tomato	Australia	Un
(*Zeugodacus*) *cucurbitae* Coquillett	melon fly	cucurbits, solanaceous, >125 hosts	S.E. Asia, Taiwan, Japan, Hawaii	CL
(*Bactrocera*) *curvipennis* Froggatt		citrus	New Caledonia	CL
(*Paradacus*) *decipiens* Drew		cucurbits	New Britain	Un
(*Dacus*) *demmerezi* Bezzi		cucurbits	Madagascar, Reunion	Un
(*Dacus*) *disjunctus* Bezzi		cucurbits	Nigeria	Un
(*Hemigymnodacus*) *diversus* Coquillett		cucurbits, mango	India, Pakistan, Burma	CL

Table 5.1 (continued). *Dacus* Fruit Flies That are Pests of Agriculture*

Dacus spp.	Common name	Hosts	Location	Male lure[1]
(*Bactrocera*) *dorsalis* Hendel	oriental fruit fly	>173 species of fruits	S.E. Asia, Indonesia, Philippines, Hawaii	ME
(*Bactrocera*) *fascialis* Coquillett		citrus, mango, guava, peach, tomato, capsicum	Tonga	CL
(*Bactrocera*) *frauenfeldi* Schiner		mango, guava, breadfruit	Australia, Solomons	CL
(*Didacus*) *frontalis* (Becker)	"melon fly"	cucurbits	Egypt, S. Africa, Arabia, Cape Verde	CL
(*Bactrocera*) *halfordiae* Tryon		citrus, loquat	Australia	Un
(*Afrodacus*) *jarvisi* (Tryon)		citrus, mango, papaya, deciduous fruits	Australia	Un
(*Bactrocera*) *kirki* Froggatt		peach, mango, guava	Samoa, Tahiti	CL
(*Bactrocera*) *latifrons* (Hendel)	Malaysian fruit fly	tomato, egg plant, peppers	S.E. Asia, Taiwan, Hawaii	α-ionol
(*Bactrocera*) *mayi* Hardy		apricot	Australia	ME
(*Bactrocera*) *melanotus* Coquillett		citrus, mango, guava	Cook Islands	ME
(*Bactrocera*) *melas* (Perkins & May)		citrus, guava, date, deciduous fruits	Australia	CL
(*Bactrocera*) *musae* (Tryon)	banana fruit fly	banana, tomato, papaya, guava, chillies	Australia, New Guinea	ME
(*Bactrocera*) *mutabilis* (May)		kumquat	Australia	Un
(*Bactrocera*) *neohumeralis* Hardy		citrus, deciduous, tomato, coffee	Australia	CL
(*Bactrocera*) *nigrotibialis* (Perkins)		coffee	S.E. Asia	CL
(*Bactrocera*) *occipitalis* (Bezzi)		carambola fruit	Malaysia, Philippines	CL
(*Daculus*) *oleae* (Gmelin)	olive fly	olive	Mediterranean, Africa, India	CL

Table 5.1 (continued). *Dacus* Fruit Flies That are Pests of Agriculture*

Dacus spp.	Common name	Hosts	Location	Male lure[1]
(*Bactrocera*) *passiflorae* Froggatt		citrus, mango, grenadilla, passion fruit	Fiji, Tonga	CL
(*Bactrocera*) *pedestris* (Bezzi)		carambola fruit	Malaysia	Un
(*Bactrocera*) *psidii* (Froggatt)		citrus, mango, guava, grenadilla	New Caledonia	CL
(*Dacus*) *punctatifrons* Karsh		cucurbits	Mauritius, Zimbabwe	CL
(*Zeugodacus*) *scutellaris* Bezzi		cucurbits	India, Pakistan	Un
(*Zeugodacus*) *tau* (Walker)		cucurbits, mango, clove	Malaysia, Philippines	CL
(*Dacus*) *telefaireae* (Bezzi)		cucurbits	Zimbabwe	CL
(*Bactrocera*) *trivialis* Drew		peach, guava	New Guinea	CL
(*Bactrocera*) *tryoni* Froggatt	Queensland fruit fly	>106 species of fruits	Australia	CL
(*Bactrocera*) *umbrosus* F.	jackfruit fly	breadfruit, citrus, mango	Australia, Malaysia, Indonesia, Philippines	ME
(*Didacus*) *vertebratus* Bezzi	"melon fly"	cucurbits	Nigeria, Zimbabwe	methyl p-hydroxy benzoate
(*Notodacus*) *xanthodes* (Broun)		citrus guava, grenadilla, tomato, papaya, pineapple	Fiji, Somoa, Tonga	ME
(*Bactrocera*) *zonatus* (Saunders)	peach fruit fly	citrus, mango, tomato, guava, deciduous fruits, custard apple	India, Pakistan, S.E. Asia	ME

[1] CL = cue-lure, ME = methyl eugenol, Un = unknown

* Data from Drew et al. (1978), Drew (1989), Carey & Dowell (1989).

Table 5.2. Ceratitis Male Fruit Flies Attracted to Terpineol Acetate*

Ceratitis spp.
(Acropteromma) munroanum Bezzi
(Carpophthoromyia) dimidiata (Wiedeman)
(Ceratalaspis) cosyra (Walker)
(Ceratalaspis) punctata (Wiedeman)
(Ceratalaspis) quinaria (Bezzi)
(Hoplaphomyia) divaricata Munro
(Pardalaspis) bremmi Guerin-Meneville
(Pardalaspis) lobata Munro
(Pardalaspis) pedestris Bezzi
(Pterandrus) capitata Wiedeman
(Pterandrus) cornutus Bezzi
(Pterandrus) podocarpi Bezzi
(Pterandrus) rosa Kharsh
(Pterandrus) rubivorus Coquillet

* Data from Ripley & Hepburn (1935), Hancock (1985a).

Host plant odors acting as kairomones are particularly significant in the ecology of fruit flies, attracting them to plants that originally were scattered through dense tropical forest canopy. Some of these kairomones have great evolutionary and behavioral significance and this forms the subject matter of this chapter.

II. CERATITINAE FRUIT FLIES

The fruit flies of the subfamily Ceratitinae comprise about 100 species and are thought to have evolved in Africa. At least six species of the genus *Ceratitis* are known pests, and the Mediterranean fruit fly *C. capitata* has been distributed to Mediterranean type climates in Africa, Europe, the Middle East, Australia, South America, Central America, Mexico, and Hawaii, where it is a notorious pest because of its great host range. The Natal fruit fly *C. rosa* is an important pest of South Africa. Other species are listed in Table 5.2.

A. Mediterranean Fruit Fly

This fruit fly, *Ceratitis capitata* (Wiedemann), is one of the world's most destructive insects. The Medfly is a native of tropical West Africa, was found in Spain in 1842 and subsequently spread into France, Italy, Greece, and the Middle East. Commerce with infested fruits introduced the Medfly into Australia in 1893, South America in 1901, Hawaii in

1907, and into Costa Rica in 1935. From there it has moved south through Central America into western South America and north into southern Mexico by 1972. During favorable mild winters, the Medfly invades central Europe and, during an outbreak in 1955, damaged peaches in northern Italy, Switzerland, France, and Germany.

The Medfly was first recorded in the continental United States in Florida in 1929, and was eradicated by arsenical-molasses bait sprays and mass destruction of infested fruits at a cost of $7 million. A subsequent reinvasion occurred in 1956 when it infested 28 counties over 1000 square miles, and was eradicated by malathion-protein bait at a cost of $10 million. Additional reinvasions occurred in Florida in 1962–1963 and in 1983; and in Texas in 1966. The Medfly invaded California in 1975, infesting an area of 100 square miles, and was eradicated at a cost of $1.5 million. The most extensive invasion of California occurred in 1980, and the Medfly was finally eradicated from a total area of 4,157 square miles using malathion - corn protein hydrolysate bait applied by aircraft at a rate of 12 fluid ounces per acre at weekly intervals. In addition, fly reproduction was curtailed by the release of 4,345,400,000 sterile male flies over the period of 7–14–80 to 7–16–81. The total cost of this eradication program was estimated at $100 million, with an additional $100 million in lost markets for the fruit industry (Rohwer 1987). Presently California is undergoing yet another invasion of the Medfly, and there is speculation that the pest has become endemic. The total cost of eradication, control, and crop losses from Medfly infestations in California over the period of 1981–1987 has been estimated at $350 million. Should the Medfly become endemic, it has been estimated that the annual cost of control efforts using malathion-protein bait would be $382 million, and that quarantine treatments could cost an additional $100 million annually (Citrograph 1990).

1. Life History, Appearance, and Habits

The Medfly adult is about the size of a house fly, but has a shiny black thorax, a yellowish abdomen with two silver cross bands, and wings banded with yellow-orange. The adult flies have sponging mouthparts and feed on honeydew excreted by aphids, mealy bugs and scale insects. The males become sexually mature within a few days after emergence, and are attracted to the volatiles emanating from ripening fruits where they rendezvous with the females and mate at temperatures of $> 18°$ C, which is also the critical temperature for oviposition. The females must have proteinaceous food to produce eggs, and become gravid after about one week at summer temperatures. They generally remain in foliage near areas where they emerged from their pupal cases in the soil. The female fly uses her sharp ovipositor to puncture the dermis of ripening fruits, making a cavity where an average of 4 to 10 eggs are deposited. A single female can oviposit up to 40 eggs a day, and produces as many as 1000 eggs over a lifetime of about 60 days. Maximum longevity is 230 to 315 days.

The glistening white eggs in the fruit hatch in 2 to 3 days at 26° C, but do not develop below 12° C, forming small cream colored, legless maggots that bore through the pulp of the fruit. Over 100 larvae develop in a large fruit. Development through three instars requires about 6 to 11 days at 24–26° C, and 24–50 days at 11–17° C, with no growth at 10° C. Mature larvae emerge from the fruit and enter the soil, where they pupate over a period of about 9 to 11 days at 24° C. Under cool conditions, pupal development may require 60 days, and development ceases at 10° C. The total life cycle from egg to adult requires about one month, and under tropical and subtropical climates there are 11 to 13 generations a year, but only 2 to 3 in temperate climates (Hagen et al. 1981).

Host Range. The Medfly has been recorded as developing in 253 species of fruits, nuts, and vegetables, of which the majority are of tropical origin. According to Hagen et al. (1981) 40 of these host plants are heavily or generally infested, and those of especial importance to California agriculture include citrus, loquat, cherry, apricot, peach, plum, quince, pear, apple, fig, persimmon, and guava. Other plants that are sometimes infested include avocado, bell pepper, tomato, walnut, grape, and cotton. In Hawaii, coffee is infested.

B. Attractants for Ceratitinae

1. Kerosene

Kerosene is an effective lure for *Ceratitis capitata* and plays an anomalous role in this discussion of plant kairomones. According to Severin & Severin (1913), the Devenish's in Guilford, Australia first observed the attraction of the Medfly to kerosene in 1907, and caught about 2000 in their citrus orchard during one week of trapping. The Severins (1913) described a number of extensive efforts to trap the Medfly with kerosene in Australia over the period of 1907–1910, and reported their own experiments in Hawaii in 1912 where 10 traps caught 10,293 flies, of which only 36 were females. A heavy infestation of orchard fruit resulted despite the extensive trapping, and the kerosene lure was considered a failure for Medfly control. Howlett (1915) who had just discovered the attraction of

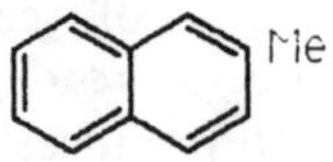

2-methylnaphthalene

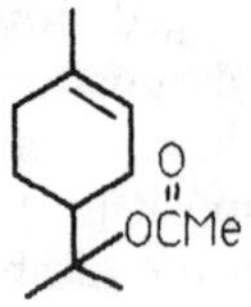

α-terpineol acetate

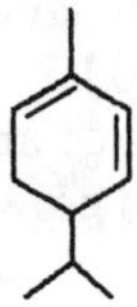

α-phellandrene

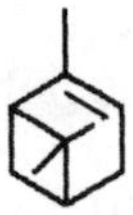

α-pinene

citronella oil and its active component methyl eugenol to male *Dacus zonatus*, the peach fruit fly (as discussed later), stated that the attraction of the male Medfly to kerosene is apparently an analogous case.

The attraction of male Medfly to kerosene remains an olfactory conundrum. In the extensive survey of fruit fly attractants made by Beroza & Green (1963), 2-methylnaphthalene, a minor constituent of kerosene, was identified as an attractant for male Medfly, although the isomer 1-methylnaphthalene was of very low attractivity. Further olfactometer and field testing showed that 2-methylnaphthalene was the only active attractant chemical in kerosene, although it was less attractive than the newer synthetic attractant trimedlure (Teranishi et al. 1987).

2. Terpineol Acetate

Terpineol acetate or α,α-4-trimethyl-3-cyclohexene-1-methylacetate, was found by Ripley & Hepburn (1935) to be an effective attractant for a number of species of Ceratitinae fruit flies, including *C. capitata*, the Mediterranean fruit fly, and *C. rosa*, the Natal fruit fly. No other genera of fruit flies were attracted. Olfactory response to terpineol acetate is widespread among male African Ceratitinae (Ripley & Hepburn 1935, Hancock 1985a) (Table 5.2). For example, 150 traps in a citrus orchard, baited with terpineol acetate, caught 19,800 male *C. rosa* (Ripley & Hepburn 1935). Terpineol acetate is a plant kairomone found, for example, in cejeput oil, petigrain oil, and long-leaf oil.

The 1-methylcyclohexene moiety of α-terpineol acetate is also found in other effective Medfly attractants, e.g. the naturally occurring α-co-α-copaene 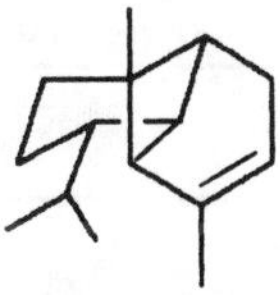 α-ylangene

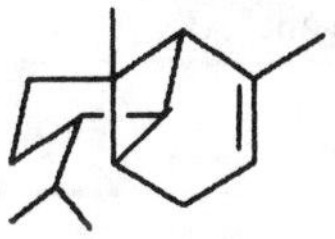

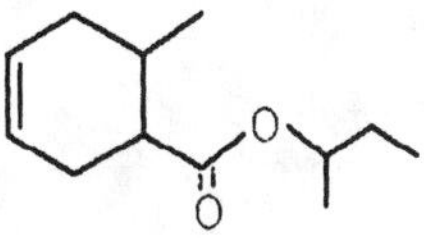

siglure 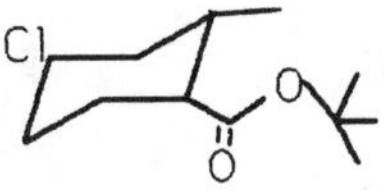 trimedlure

paene and α-ylangene. Related structures occur in the synthetic parakairomones siglure, medlure, and trimedlure. These lures are discussed under parakairomones. It is apparent that the 1-methylcyclohexene moiety is essential for male antennal receptor complimentarity in the Ceratitinae (Guiotto et al. 1972).

3. Alpha-*Copaene and* Alpha-*Ylangene*

Angelica root oil from *Archangelica officinalis* was shown to be attractive to male *C. rosa*, the Natal fruit fly, by Ripley & Hepburn (1935). Additional study by Steiner et al. (1957) showed that angelica seed oil was highly attractive to male *C. capitata*, the Mediterranean fruit fly. Field testing in Hawaii showed that 6 to 10 traps baited with 0.5 ml of angelica seed oil caught 25,500 Medfly males weekly, as compared to 2600 males caught in traps baited with protein hydrolysate-ammonium chloride. The angelica seed oil was used to bait about 50,000 traps in Florida during the 1956 Medfly eradication program (Steiner et al. 1957).

The active kairomone components of angelica seed oil have been identified as the tricyclic sesquiterpene α-copaene, or [1R-(1α,2α,6α,7α,8α)]-1,3-dimethyl-8-(1-isopropyl)-tricyclo-[4.4.0.$0^{2,7}$]-dec-3-ene (bp 246–251°C), and its stereoisomer α-ylangene, or [1S,2R,6R,7R,8S]-(+)-8-isopropyl-1,3-dimethyl-tricyclo-[4.4.0.$0^{2,7}$]-dec-3-ene (Guiotto et al. 1972).

Alpha-copaene is widely distributed in plants, chiefly as an ingredient of essential oils (Table 5.3) (Teranishi et al. 1987, Warthen & McInnis 1989). *Alpha*-ylangene is present in oil of *Schizandra chinensis* and in ylang-ylang, *Canaga odorata* (Warthen & McInnis 1989), and is probably

Table 5.3. Plant Sources of α-Copaene and α-Ylangene*

Plant species	Common name	Family
Ananas comosus	pineapple	Bromeliaceae
Archangelica officinalis	angelica	Umbelliferae
Canangium odoratum	ylang-ylang	Anonaceae
Cedrella toona		
Chloranthus		
Citrus aurantium	orange	Rutaceae
Copaifera	copaiba	Leguminaceae
Cyperus articulatus		Cyperaceae
Ficus retusa	fig	Moraceae
Gossypium hirsutum	cotton	Malvaceae
Heracleum dissectum		Umbelliferae
Humulus lupulus		
Lippia nodiflora		Verbenaceae
Litchi chinensis	litchi	Sapindaceae
Mentha piperita	mint	Labiateae
Phyllocladus trichomonoides		Podocarpaceae
Piper cubeba	peppermint	Piperaceae
Psidium guajava	guava	Myrtaceae
Salvia officinalis	sage	Lamiaceae
Sindora supra		Leguminoseae
Siparuna guianensis		Monimiaceae
Triticum aestivum	wheat	Graminaceae
Yucca gloriosa	yucca	Lilliaceae
Zea mays	corn	Graminaceae

* Data from Teranishi et al. (1987), Warthen & McInnis (1989).

present in many of the sources of copaene. Both α-copaene and α-ylangene are present in angelica seed oil at 0.5–1.0%. The pure compounds are highly attractive to the male Medfly at dilutions of 1×10^{-6} g per l in acetone, as compared to angelica oil, which is about equally attractive at 1×10^{-2} g per l (Guiotto et al. 1972). According to Warthen & McInnis (1989) α-copaene is slightly more attractive than α-ylangene. The α-copaene from angelica seed oil has the (+)-configuration and is considerably more attractive than (−)-α-copaene from copaiba and cubeb oils. Orange oil, however, contains (+)-α-copaene and is highly attractive (Teranishi et al. 1987). This isomer is stated to be several times more attractive than the parakairomone trimedlure.

Total syntheses for the copaenes and ylangenes have been devised by Corey & Watt (1973), but the costs of the synthetic kairomones would be very high and natural sources of these *Ceratitis* spp. attractants will be relied upon.

Table 5.4. Attractivity of Ester Analogues of *cis*-4-Chloro-*trans*-2-Methylcyclohexane Carboxylate to Male Mediterranean Fruit Flies*

Ester	Mean catch after 1 day
methyl	1.6ij
ethyl	33.3ef
propyl	39.8ef
butyl	6.6hij
tert-butyl (trimedlure ester C_2)	93.9ab
1-methylbutyl	81.8b
2-methylbutyl	5.2ij
1-methylpropyl	117.9a
2-methylpropyl	22.2fg
1-ethylpropyl	66.9bcd
1,1-dimethylpropyl	122.7a
1,2-dimethylpropyl	27.6fg
2,2-dimethylpropyl	2.5ij

* Data from McGovern & Cunningham (1988).

4. Parakairomones for Ceratitis capitata

The U.S. Department of Agriculture has made extensive studies of synthetic cyclohexene and cyclohexane esters related to terpineol acetate and α-copaene as lures for the male Medfly (Beroza & Green 1963). The aliphatic esters of 6-methyl-3-cyclohexene-1-carboxylic acid are all effective attractants, and attraction as measured in the Gow olfactometer was in the order ethyl > allyl > propyl = isopropyl = isobutyl, sec-butyl > butyl = pentyl = isopentyl > methyl (Gertler et al. 1958). *Sec*-butyl 6-methyl-3-cyclohexene-1-carboxylate was developed as a practical lure, siglure; and was widely used in the 1962–1963 Florida eradication program for the Medfly. In plastic traps, siglure was stated to be about 0.16 times as active as angelica seed oil containing 0.5–1.0% α-copaene (Steiner et al. 1958). In the Gow olfactometer the *cis*-isomer of siglure is about 0.25 times as active as the *trans*-isomer, suggesting that the *trans*-configuration most closely approximates the stereochemistry of α-copaene.

Medlure and trimedlure. Further study of synthetic Medfly lures (Beroza et al. 1960) showed that reaction of the cyclohexene double bond with HCl, to produce the corresponding chloro-methylcyclohexane carboxylic acid esters, produced more attractive Medfly lures. The order of effectiveness in laboratory lure evaluations with various esters of 4-chloro-*cis*-*trans*-2-methylcyclohexane carboxylic acid, the most attractive isomer (B2 of trimedlure,) is shown in Table 5.4. In 4 week field tests, the number of male Medfly trapped with *tert*-butyl 4 (or 5)-chloro-2-methylcyclohexane carboxylate (trimedlure) was 16,062; with the *sec*-butyl ester (medlure) was 11,396, and with siglure at 2 times dosage and double appli-

Table 5.5. Attraction of Male Mediterranean Fruit Flies to the Eight *cis-trans* Isomers of Trimedlure*

Substituted *tert*-butyl 2-methyl-cyclohexane carboxylate	Isomer	Mean catch after 1 day I	II
cis-4-chloro-*trans*	(trimedlure C)	143a	158a
trans-5-chloro-*trans*	(A)	99ab	43c
cis-5-chloro-*trans*	(B_1)	49cd	16d
trans-4-chloro-*trans*	(B_2)	11e	1e
cis-5-chloro-*cis*	(Y)	78bc	40c
trans-5-chloro-*cis*	(V)	28de	10d
cis-4-chloro-*cis*	(X)	21de	2e
trans-4-chloro-*cis*	(W)	153	1e

* Data from McGovern et al. (1990).

cation was 6,119. It was noted that while the Medfly catches with medlure and trimedlure were almost entirely of male flies, the Medfly male did not feed as readily on the synthetic parakairomones as on angelica seed oil, but congregated around the treated wicks.

Trimedlure consists of 90–95% of the four possible *trans*-isomers (with regard to *tert*-butyl-cyclohexane carboxylate) (McGovern et al. 1990). The commercial trimedlure contains 70–75% of the *tert*-butyl *trans*-5-chloro-*trans*-2-methylcyclohexane carboxylate and the *tert*-butyl *cis*-4-chloro-*trans*-2-methylcyclohexane carboxylate, which are the two most attractive isomers, as shown in Table 5.5. The order of attraction for the *trans*-isomers is: *cis*-4-chloro-*trans* > *trans*-5-chloro-*trans* > *cis*-5-chloro-*trans* > *trans*-4-chloro-*trans* (Table 5.5). The *tert*-butyl *cis*-5-chloro-*cis*-2-methylcyclohexane carboxylate, present to about 5%, was attractive (Table 5.5), but the other *cis*-isomers were unattractive. The most favorable configuration about the cyclohexane ring for attraction of male Medfly requires an axial Cl atom and *trans*-diequatorial methyl and *tert*-butyl ester groups (McGovern et al. 1990).

These studies clearly demonstrate that the overall spatial arrangement of the lure molecule determines its interaction with the antennal kairomone receptor. The data from Table 5.5 can be summarized as:

1) *trans-tert*-butyl carboxylates are more attractive than *cis*-isomers.
2) the presence of all three substitutions of the cyclohexane ring, (a) 1-$C(O)\text{-}OC(CH_3)_3$, (b) 2-CH_3, and (c) 4-(or5)-Cl, are essential for highest activity.
3) there are inter-molecular interactions between the *tert*-butyl carboxylate group and the 2-CH_3 and 4 (or 5)-Cl groups.

4) epimers with axial Cl atoms are more attractive than those with equatorial Cl atoms.
5) the most attractive molecular combination has axial-Cl and *trans*-diequatorial CH_3 and $C(O)\text{-}OC(CH_3)_3$ groups.

Sonnett et al. (1984) have reported that for isomer C (Table 5.5), (*cis*-4-chloro-*trans*-ester) the 1S, 2S, 4R enantiomer was very attractive and the 1R, 2R, 4S enantiomer was unattractive. However, the two corresponding enantiomers of isomer A (*trans*-5-chloro-*trans*-ester) were equally attractive.

In an evaluation of the attractivity of aliphatic ester analogues of *tert*-butyl *cis*-4-chloro-*trans*-2-methylcyclohexyl carboxylate (Table 5.5, isomer B2) (McGovern & Cunningham 1988), the structure of the alcohol moiety had a marked effect on the attractivity of the ester. All of the most attractive analogues were esters of bulky, branched C_5 or C_6 alcohols, with methyl or ethyl groups either in the 1-position or in the 1,1-dimethyl branching. Esters of primary alcohols were poor attractants (Table 5.4). These results support the hypothesis that the overall molecular structure of the trimedlure-type parakairomone is critical for maximum attractivity. This suggests that the ester moiety of these molecules is involved in complimentarity to the α-copaene receptor.

5. Structure-Activity Summary for Ceratitis capitata

The identification of α-copaene and α-ylangene as kairomone attractants for male Medfly at dilutions of 1×10^{-6} g per l, compared to α-pinene at 10 g per l (Guiotto et al. 1972) points to the importance of the 1-methyl-2-cyclohexene moiety as a complimentary entity to the male antennal receptor of *C. capitata.* This moiety is present in both α-copaene and α-ylangene, and is also found in the attractants α-pinene and terpineol acetate. An isomeric variant is present in the parakairomone siglure, and the saturated cyclohexane variant is present in medlure and trimedlure. The terpenoid α-phellandrene, or 2-methyl-5-isopropyl-1,3-hexadiene, with two $C = C$ bonds, is also attractive to the Medfly (Guiotto et al. 1972) β-caryophyllene was stated to be an attractant for *Ceratitis* spp. (Ripley & Hepburn 1935, Beroza & Green 1963), although this attraction is actually the result of α-copaene impurity (Warthen & McInnis (1989).

Inspection of Dreiding molecular models shows that there is not only a marked similarity between the methylcyclohexene moieties of the natural kairomones α-copaene, α-ylangene, α-terpineol acetate, and α-pinene; but also that the positions of the propyl side chains in both α-copaene and α-ylangene are approximated sterically to the *O*-isopropyl, *O-sec-*

butyl, and *O-tert*-butyl esters of the siglure, medlure, and trimedlure types (Guiotto et al. 1972, Teranishi et al. 1987)

The only data we have found for the relative attractancy of α-terpineol acetate is the rating of 2 by Beroza & Green (1963), as compared to 3 for siglure, trimedlure, and angelica seed oil. From this fragmentary information it appears that α-copaene is very likely the original plant kairomone with which *C. capitata* and other Ceratitinae coevolved, and that complete and rapid depolarization of the male antennal receptor in these Ceratitinae requires both the methylcyclohexene ring moiety and a sterically appropriate isopropyl group in a fixed position on an adjacent fused cyclohexane ring. Only a part of this complete structure is represented by either α-terpineol acetate or the synthetic parakairomones.

The substantial differences in attractivity of (+) and (−)-α-copaene, the *cis*- and *trans*-isomers of siglure, and the eight isomers of trimedlure, indicate the extreme degree of molecular complimentarity between kairomone and antennal receptor required for maximum depolarization. This emphasizes the profound evolutionary association between plants producing α-copaene and ancestral Ceratitinae.

III. DACINAE FRUIT FLIES

Fruit flies of the genus *Dacus* (subfamily Dacinae) number about 700 described species that were originally restricted to the tropics and subtropics of the Old World. The preponderance of the species are monophagous or stenophagous breeding in host plants of rain forests and, many species are concentrated in the Cucurbitaceae, Asclepeadaceae, and Passifloraceae. However, the hosts of only about one-quarter of the species are known (Munro 1984, Drew 1989, Metcalf 1990). In contrast, more than 40 *Dacus* spp. identified in Table 5.1 breed in the fruits of cultivars and are serious economic pests, or have the potential to become so if imported into areas of tropical and subtropical horticulture.

A. Life Histories, Appearance, and Habits

The developmental patterns of the polyphagous Dacinae, e.g. *D. cucurbitae*, *D. dorsalis*, and *D. tryoni* are very similar (Fletcher 1987). The females insert slender, white, glistening eggs in groups of 10–50 under the cuticle of fruits through a long, protrusible ovipositor. Maximal egg production occurs at 25–30° C, and each female may produce 1000–1500 eggs. The eggs hatch in 1 to 2 days, and the headless maggots pass through 3 larval instars over a period of 7 to 8 days at 25° C. The last instar larvae are 10–11 mm long, and escape from the infested fruits and drop to the ground where they pupate in yellowish to brown puparia. Pupariation occurs over 10–11 days, and the females become sexually mature within 2–5 days. The temperature-development rate curves of the several species are of the same general shape, with a low temperature threshold of 6–9° C, maximum development at 26–30° C, and a developmental decrease at 30° C (Fletcher 1987). Although the generation time of the tropical species is largely determined by temperature, the fruit variety, ripeness, and overcrowding of larvae can have significant effects. Maximum productivity of *D. dorsalis* in its favorite host, the guava, occurs at larval densities of 0.8 to 1.6 larvae per g of fruit (Bateman 1972).

According to Fletcher (1987) the polyphagous Dacinae produce > 1000 eggs per female, oligophagous species produce 400–600 eggs, and the monophagous olive fly *D. oleae* produces about 300 eggs laying each in a single olive fruit.

The oriental fruit fly, *Dacus dorsalis*, is indigenous to S.E. Asia, Indonesia, and the Philippines and has been introduced into Taiwan, Hawaii, and other S. Pacific Islands. It was brought into Hawaii by military air transport in 1946. *D. dorsalis* has been reared from 173 plant species in 112 genera, including all the common fruits, but does not attack cucurbits. The oriental fruit fly is somewhat larger than the Mediterranean fruit fly, with a dark brown body bearing a conspicuous yellow dorsal abdominal band. The adults emerge from brown puparia after about 10 days, and following a preoviposition period of 8–12 days, the female lays from 1200–1500 eggs. The entire life cycle requires about 16 days at normal subtropical temperatures.

The oriental fruit fly has been introduced into California at least 12 times since 1966 and has been eradicated each time by the use of the male lure methyl eugenol in malathion bait spray. California crops susceptible to fruit fly infestations are valued at more than $4.5 billion annually.

The melon fly, *Dacus cucurbitae*, is indigenous to S.E. Asia, and was introduced into Hawaii from Japan in 1895. It has been reared from more than 125 host plants, and is a particular pest of melon, squash, cucumber and other cucurbits, eggplant, tomato, and beans. The adult melon fly is slightly larger than the oriental fruit fly, and has darkly patterned wings. The female inserts slender white eggs about 2 mm long under the skin of fruits in groups of 1 to 37, and these hatch in about 26 hours forming white, legless maggots. They pass through 3 instars over a period of 4 to 17 days, and when mature, about 11 mm long, drop to the ground and pupate in the soil in a yellow-brown puparium. The adults emerge after about 9 days. The melon fly has been trapped in California several times, and this species represents a severe threat to the California tomato industry (Carey & Dowell 1989).

The Queensland fruit fly, *Dacus tryoni*, is found in Australia, and is an ecological homologue of *D. dorsalis*, attacking more than 106 species of fruits. Both *D. tryoni* and *D. dorsalis* are polyphagous generalists capable of breeding in almost all fleshy fruits (Drew et al. 1978).

The olive fly, *Dacus oleae* is essentially monophagous, attacking only *Olea europa* and *O. africanus* (Oleaceae), but is immensely destructive because it inhabits the entire Mediterranean region, where olives are grown commercially.

Carey & Dowell (1989) consider the following *Dacus* spp. to be serious potential threats to California agriculture: *D. bivitattus*, *D. ciliatus*, *D. citri*, *D. correctus*, *D. cucurbitae*, *D. dorsalis*, *D. latifrons*, *D. oleae*, *D. tryoni* and *D. zonatus* (Table 5.1) Drew et al. (1978) consider the following species, which inhabit neighboring islands, to have the potential to become major pests in Australia: *D. decipiens*, *D. facialus*, *D. frauenfeldi*, *D. kirki*, *D. melanotus*, *D. passiflorae*, *D. psidii*, and *D. xanthodes* (Table 5.1).

The establishment of a major pest fruit fly in California or Florida would have far-reaching effects on subtropical horticulture because of the high value of these crops, which are exported, and because of the resultant plant quarantines established by importing countries. Fruits and vegetables for export must be disinfested by treatment with vapor heat, fumigation, or radiation, and these procedures not only increase costs by 10 to 100 percent depending upon the infesting species and the plant commodity, but are also likely to impair commodity flavor and quality. The recent removal of ethylene dibromide from the list of acceptable fumigants leaves only methyl bromide and phosphine as viable treatments to kill the larvae in infested fruits (Carey & Dowell 1989). Eradication programs are therefore mandated by the Federal Government to ameliorate the adverse effects of new fruit fly infestations on the U.S. agricultural economy. A recent U.S.D.A. estimate puts the cost of fruit fly importation and infestation into California at $800 million annually (Honolulu Star Bulletin 1990).

B. Attractants for Dacinae

Howlett's (1912) observation that oil of citronella (*Cymbopogon nardus*, Andropogonaceae) was attractive to the male fruit flies, *Dacus diversus* and *D. zonatus*, was a landmark in the study of the chemical ecology of insects. Further investigation (Howlett 1915) showed that the attractive

Table 5.6. Cue-lure and Raspberry Ketone as Attractants for Male Melon Flies*

	Mean number of flies caught after indicated weeks					
	1	2	4	8	16	32
cue-lure (100%)	1076	498	1161	1714	1278	2164
raspberry ketone (100%)	991	599	1555	2544	1098	1242

* Data from Keiser et al. (1973).

component was the phenyl propanoid methyl eugenol, or 3,4-dimethoxy-1-allylbenzene, which was also extremely attractive to *D. dorsalis* (*D. ferruginius*). It is noteworthy that this identification of the chemistry of a plant kairomone for insects antedated the chemical characterization of the first insect sex pheromone, bombykol, by more than 40 years. Howlett's discovery was almost forgotten until the oriental fruit fly was discovered in Hawaii in 1946, and its dramatic spread and increase initiated a search for lures, and the effectiveness of methyl eugenol was rediscovered (Steiner 1952).

The melon fly, brought to Hawaii about 1885 (Hardy 1979), became a ubiquitous pest of melons, squash, egg plant, and tomatoes. Thousands of chemicals were screened as lures for the oriental, melon, and Mediterranean fruit flies (Beroza & Green 1963) and anisyl acetone, or 4-(*p*-methoxyphenyl)-2-butanone, was discovered to be an effective attractant for the melon fly (Barthel et al. 1957). A derivative cue-lure, or 4-(*p*-acetoxyphenyl)-2-butanone, was found to be a more effective lure for this species (Beroza et al. 1960). Cue-lure has not been isolated as a natural product but is rapidly hydrolyzed to form 4-(*p*-hydroxyphenyl)-2-butanone, rheosmin, or raspberry ketone. This kairomone was first isolated from Chinese rhubarb *Rheum palmatum* (Polygonaceae) where it occurs as the *para*-glucoside (Bauer et al. 1955). The free phenol, raspberry ketone was found in the raspberries *Rubis idaeas* and *R. sligosis* by Schinz & Seidel (1961), and also occurs in the juice of cranberries, *Vaccineum oxyococceus* and *V. macrocarpa* (Ericaceae), and in the aerial parts of *Scutellaria rivalaris* (Lamiaceae) (Lin & Chen 1984). Raspberry ketone was developed about 1959 as Willisons' lure for the male Queensland fruit fly, *D. tryoni* (Drew 1974). Raspberry ketone is a very effective lure for the melon fly (Table 5.6) but its much lower release rate, about 0.05 times that of the para-kairomone cue-lure, makes cue-lure more efficient for long-range attraction, although raspberry ketone is much more persistent. Since cue-lure rapidly forms rasberry ketone, which is the basic fruit fly attractant, we will discuss cue-lure/raspberry ketone as a single entity.

1. Attraction of Male Dacinae to Plants

Methyl eugenol is a phenyl propanoid widely distributed in plants, and male Dacinae have been observed feeding at the leaves, blossoms, and fruits of at least 10 families of plants (Table 5.7). This plant kairomone causes the male fruit flies to congregate and to feed compulsively (Howlett 1915, Kawano et al. 1968, Fletcher et al. 1975, Shah & Patel 1976). Nishida et al. (1988) have shown that wild *D. dorsalis* males in Malaysia accumulate up to 10 μg of the methyl eugenol metabolite 2-allyl-4,5-dimethoxyphenol in rectal glands, where it is released into the air at dusk, and is highly attractive to conspecific males but not to females. They suggest that this methyl eugenol metabolite promotes male-male interactions and may also act as an allomone to deter predators.

Dacus species responsive to cue-lure/raspberry ketone also congregate on certain plants (Table 5.7). *Dendrobium superbum* (Orchidaceae), which contains phenyl-2-butanone and possibly raspberry ketone, attracts male *D. cucurbitae* (Flath & Ohinata 1982). *Spathaphyllium cannaefolium* (Araceae), the fruit fly plant, attracts several species and contains benzyl acetate and 4-methoxybenzyl acetate, which are known lures for both *D. dorsalis* and *D. cucurbitae* (Metcalf et al. 1986).

2. Dacinae Kairomone Responses as Evolutionary Phenomena

World-wide the Dacinae are astonishingly responsive to kairomone lures. Hardy (1979) estimated that at least 90% of the species is strongly attracted to either methyl eugenol or to cue-lure/raspberry ketone. Very extensive trapping experiments conducted by Drew and colleagues (Drew 1974, 1989, Drew & Hooper 1981) to determine endemic species in Australia and the South Pacific islands responding to either lure, have shown that the kairomone lure response has important evolutionary and systematic implications. The two lures have become important tools for collecting endemic species of Dacinae and for monitoring incipient infestations of fruit flies throughout the tropics and subtropics.

A summation of the attractive responses of male Dacinae to the kairomone lures has shown that at least 176 species respond to cue-lure/raspberry ketone, and 58 species respond to methyl eugenol (Drew 1974, 1989, Drew & Hooper 1981, Hancock 1985b). Of the 46 Dacus species that are agricultural pests (Table 5.1), 24 respond to cue-lure/raspberry ketone and 8 to methyl eugenol. No species has ever been recorded as responding to both lures. The possible kairomone lure responses of several hundred species of Dacinae apparently have not been determined. However, at least 31 species, known to be present in areas where lure trapping

Table 5.7. Plants attractive to male Dacinae fruit flies

Plant family	Plant species	*Dacus* spp.	Kairomone lure	References
Anacardiaceae	*Mangifera indica* L.	*correctus*	Methyl eugenol	Howlett (1915)
		diversus		
Araceae	*Colocasia antiquorum* (Schott)	*dorsalis*	Methyl eugenol	Howlett (1915)
		zonatus		
Bromeliaceae	*Vriesea heliconiodes* (Humb.)	*dorsalis*	Methyl eugenol	Mitchell (1965)
Caricaceae	*Carica papaya* L.	*diversus*	Methyl eugenol	Howlett (1915)
Labiatae	*Ocimum basilicum* L.	*correctus*	Methyl eugenol	Shah & Patel (1976)
	O. sanctum (L.)	*dorsalis*	Methyl eugenol	Tan (1983)
Lecythidaceae	*Couroupita guianensis* Aublet	*dorsalis*	Methyl eugenol	Kawano et al. (1968)
Leguminosae	*Cassia fistula* L.	*dorsalis*	Methyl eugenol	Kawano et al. (1968)
Liliaceae	*Spathiphyllum cannaefolium*	*cacuminatus*	Benzyl acetate	Lewis et al. (1988)
		dorsalis		
		musae		
		occipitalis		
Myrtaceae	*Pimenta racemosa* (Miller)	*caudatus*	Methyl eugenol	Shah & Patel (1976)
	Syzygium cumini (L.)	*diversus*		
	S. aromaticum (L.)	*dorsalis*		
		zonatus		
Orchidaceae	*Dendrobium superbum* Rchb.	*cucurbitae*	4-Phenyl-2-butanone	Flath & Ohinata (1982)
Piutaceae	*Pelia anisata* Mann	*dorsalis*	Methyl eugenol	Fletcher et al. (1975)
	Zieria smithii Andrews	*cacuminatus*		
Saxafragaceae	*Brexia madagascariensis* Lamark	*dorsalis*	Methyl eugenol	Mitchell (1965)

Reprinted with permission from Metcalf (1990).

Table 5.8. Dacinae Fruit Flies Attracted to Kairomone Lures.

Geographic area	Described	Dacinae spp. attracted to Cue-lure/ raspberry ketone	Methyl eugenol	Reference
Africa	196	24	0	Munro (1984), Hancock (1985b)
Australia	78	39	16	Drew & Hooper (1981)
India/Pakistan	21	6	4	Kapoor & Agerwal (1983)
New Guinea	48	17	6	Drew (1974)
Philippines	44	4	4	Hardy (1974)
South Pacific	108	56	23	Drew (1974)
Thailand	52	7	7	Hardy (1973)

took place, were not collected by either methyl eugenol or raspberry ketone and are considered non-responsive (Drew & Hooper 1981, Hancock 1985b). These include at least 10 agricultural pests (Table 5.1).

The evolutionary and zoogeographic implications of the Dacinae kairomone lure responses are intriguing (Metcalf 1990). As shown in Table 5.8, of the African spp. evaluated 24 respond to cue-lure/raspberry ketone and none to methyl eugenol (Hancock 1985b). Apart from the African region, the ratio of responsiveness to cue-lure/methyl eugenol ranges from 2.5 in Australia and 3.0 in New Guinea to approximately 1 in India, Pakistan, Thailand and the Philippines. For the Oriental region, however, many more species need to be evaluated before the significance can be judged.

The two categories of Dacinae lure responses, i.e. to methyl eugenol and to cue-lure/raspberry ketone, are correlated with systematic classification based on morphological characteristics (Drew 1974, Drew & Hooper 1981). African species of Dacinae apparently respond only to cue-lure/raspberry ketone (Hancock 1985b) and of the 182 species (Munro 1984), all but 9 have fused abdominal terga and belong to the *Dacus* subgenus. Of the 290 South Pacific species, 263 have free abdominal terga (*Bactrocera* subgenus), and 94 are known to respond to cue-lure/raspberry ketone and 40 to methyl eugenol (Drew 1989). Only 27 of the species with fused terga have penetrated into the South Pacific region; 18 of these respond to cue-lure/raspberry ketone and 3 to methyl eugenol.

Specific kairomonal responses are associated with closely related complexes of morphologically similar species (Drew 1989). The methyl eugenol response is found in all 8 species of the *dorsalis* complex, all 3 species of the *musae* complex, and all 3 species of the *nigella* complex of the *Bactrocera*. The cue-lure/raspberry ketone response is found in all 8 species of the *aemula* complex, all 8 species of the *silvicola* complex, all 4 species of the *tryoni* complex, and 12 of 14 species of the *distincta* complex (Drew 1989).

This information raises interesting speculations about the zoogeographic origin of the Dacinae; Munro (1984) believing them to be native to Africa and Drew (1989) believing them to be native to New Guinea (Metcalf 1990).

3. Physiology and Ecology of Kairomone Responses

Dacinae responses to the kairomone lures methyl eugenol and raspberry ketone are qualitatively and quantitatively very similar. Both kairomones are male lures, and males of *D. dorsalis* (methyl eugenol) and *D. cucurbitae* (raspberry ketone) are rapidly attracted to nanogram quantities of the lures on filter paper in 1 ft^3 cages. The responses involve identical behavioral sequences of orientation, searching, attraction, pulsation of the mouthparts, compulsive feeding, and regurgitation (Metcalf 1990). Orientation of male flies has been observed with as little as 0.0001 μg on filter paper, and male *D. dorsalis* were attracted to ^3H methyl eugenol when the antennal deposition was 100 pg (Metcalf et al. 1979).

Responsiveness to either methyl eugenol or to raspberry ketone is associated with sexual maturity, and male *D. dorsalis*, *D. opiliae*, *D. cucurbitae*, and *D. tryoni* are unresponsive to the kairomone lures until 2–4 days after eclosion. The percentage of response then rises in a sigmoid curve reaching a maximum of 80–95% of the males after 8–14 days of age (Metcalf 1990). For the several species investigated, lure response is correlated with diurnal cycles, reaching a maximum at about 1200 hours (noon) and declining sharply at dusk.

a. Kairomone Receptor Evolution

There is strong evidence that the two groups of Dacinae distinguished by male lure responses to methyl eugenol, or to cue-lure/raspberry ketone, represent divergent evolution from a common ancestral association with plants (Metcalf 1990). Ancestral Tephritidae are thought to have fed on rotting fruits where cinnamic acid, the precursor of the phenyl propanoids, was formed from phenylalanine (Friedrich 1976) as suggested in Figure 5.1 (Metcalf 1985). *Para*-coumaric acid can be portrayed as the ancestral kairomone from which there was divergence to the phloretic acid - raspberry ketone pathway, and to the eugenol - methyl eugenol pathway (Figure 5.1). As oxygenase enzymes evolved in plants, *para*-hydroxylation was a likely first step in kairomone evolution, followed by methoxylation and further oxygenation to form 3,4-dihydroxy-, 3-methoxy-4-hydroxy-, and 3,4-dimethoxyphenyl propanoids (Hanson & Havir 1979). Thus second-

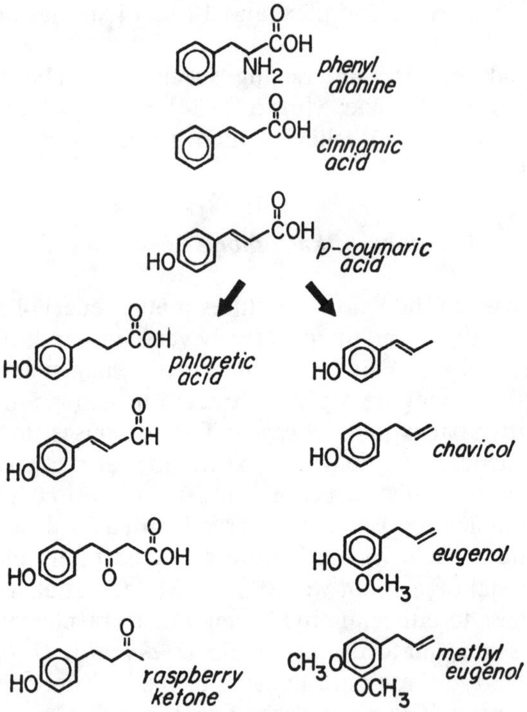

Figure 5.1. Plant evolution of phenylpropanoids from phenylalanine. Arrows indicate divergence leading to the evolution of kairomones for two distinct groups of Dacinae. Reprinted with permission from Metcalf (1975).

ary plant compounds became more lipophilic and were sequestered as plant essential oils. This speculation is supported by the positive feeding response of male *D. cucurbitae* to *p*-coumaric acid (limit of response = 1000 μg) and its much greater response to its more lipophilic ester methyl phloretate (LR = 3) (Metcalf et al. 1983). We suggest that small mutational changes in male antennal receptor sites occurred, and that these could accommodate the increasing array of lipophilic plant essential oils in the newly evolving angiosperms (Metcalf 1990). Thus *D. dorsalis* and the related Dacinae spp. responding to methyl eugenol are probably descendents of a mutant form whose antennal receptor accommodated the 3,4-dimethoxyphenyl propanoids. There is good evidence that the primary interactive site on the male antennal receptor of *D. dorsalis* is complementary to the *p*-methoxy-group of methyl eugenol, e.g. the positive response to *p*-methoxyallylbenzene (LR = 3) (Metcalf et al. 1983).

The typical male feeding and regurgitation responses of *D. vertebratus* to methyl *p*-hydroxybenzoate (Hancock 1985c) and of *D. latifrons* to α-ionol (McGovern et al. 1989) provide evidence of further receptor evolution in the Dacinae to exploit a wider range of plant kairomones.

C. Mapping the Active Sites for Kairomone Receptors

1. Dacus dorsalis Methyl Eugenol Receptor

The maximum sensitivity of attraction of caged male *D. dorsalis* to methyl eugenol residues on filter paper is about 10^{-4} μg. This is about the same order of detection as that of many male moths (Lepidoptera) to their female sex pheromones, and is evidence of an extraordinarily well developed and highly specific antennal receptor system for this plant produced kairomone.

The responsiveness of male *D. dorsalis* under standardized conditions of age, number of flies per cage, time of day, and temperature to more than 200 compounds structurally related to methyl eugenol has been quantified using the LR values (Metcalf et al. 1975, 1979, 1981, 1983). The LR was considered as positive only when the male flies were attracted to the treated filter paper, arrested, fed, and then regurgitated, as shown by the characteristic brown spots at the application site on the filter paper. From this information we can form a rather precise idea of the nature of the complementarity between the agonist kairomone and its macromolecular receptor on the male antenna (Metcalf 1990). It is evident that maximum fit of the lure to the receptor occurs with methyl eugenol, and that optimal depolarization occurs when the receptor is occupied by phenyl propanoids with an aromatic ring attached to a three atom side chain and substituted in the 3,4-positions with alkoxy groups (Figure 5.2). Reducing the length of the aliphatic side chain from C_3 to C_2 to C_1 progressively increases the LR (Figure 5.2, analogs II, III, IV, VIII) (Metcalf et al. 1981, Metcalf 1990) i.e. decreases attractivity. Molecular size and shape alone are not sufficient for receptor depolarization, as shown by the methyl eugenol isosteres, 3,4-dimethyl-1-allylbenzene and 3,4-dichloro-1-allylbenzene, which are not attractive to *D. dorsalis* at concentrations 10^6 times that of 3,4-dimethoxy-1-allylbenzene or methyl eugenol (Metcalf et al. 1975). These three molecules have almost identical molecular profiles and volumes (Figure 5.3). Bonding between the O-atoms of the methoxy groups through their unshared electron pairs and the receptor macromolecular sites is of key importance, and receptor depolarization is maximal when these methoxy groups are adjacent. The LR for *o*-dimethoxybenzene (VIII) is 1.0 μg, but this increases to 10,000 μg for *m*-dimethoxybenzene and to 100 μg for *p*-dimethoxybenzene (Metcalf

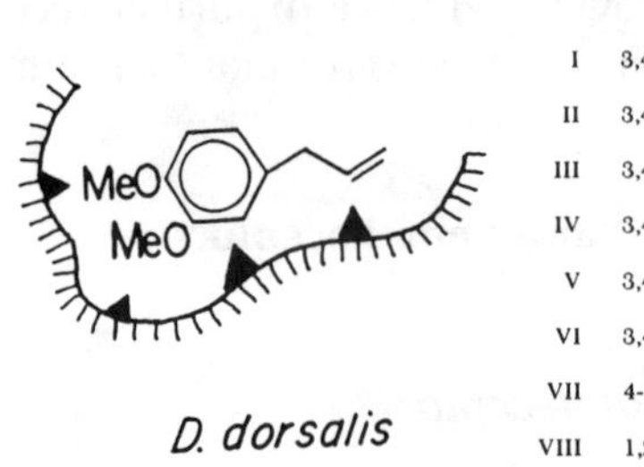

		LR — μg
I	3,4-$(MeO)_2C_6H_3CH_2CH{=}CH_2$	0.001–0.01
II	3,4-$(MeO)_2C_6H_3CH_2CH_2CH_3$	0.01
III	3,4-$(MeO)_2C_6H_3CH_2CH_3$	0.03
IV	3,4-$(MeO)_2C_6H_3CH_3$	0.3
V	3,4-$(MeO)_2C_6H_3CH_2OCH_3$	0.1
VI	3,4-$(MeO)_2C_6H_3OCH_2CH_3$	0.03
VII	4-$MeOC_6H_4CH_2CH{=}CH_2$	3.0
VIII	1,2-$(MeO)_2C_6H_4$	1.0
IX	3,4-$(MeO)_2C_6H_4OCH_3$	0.3
X	3,4-$(MeO)_2C_6H_4OCH_2CH_2CH_3$	0.1
XI	4-$MeOC_6H_4CH_2CH_2CH_3$	10

Figure 5.2. Map of suggested receptor site interactions for *D. dorsalis* to methyl eugenol together with LR values for various analogs. Reprinted with permission from Metcalf (1990).

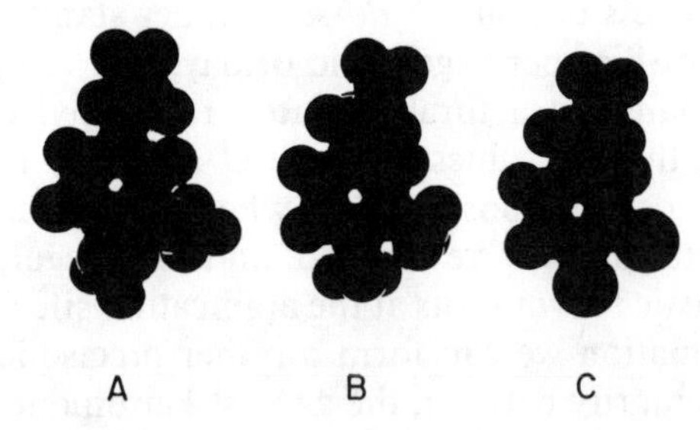

Figure 5.3. Comparative silhouettes of molecular models of 3,4-dimethoxyallylbenzene (A), 3,4-dimethylallylbenzene (B), and 3,4-dichloroallylbenzene (C). Reprinted with permission from Metcalf et al. (1975).

et al. 1979). The methoxy substituent of the phenyl ring *para* to the unsaturated allyl group is the master group, as 4-methoxy-1-allylbenzene (VII) (estragole, LR = 3.0) is appreciably attractive, but compounds with single ring methoxy groups *ortho-* or *meta-* to the unsaturated allyl side chain are unattractive (Metcalf et al. 1981). However, the presence of an additional methoxy on the aromatic ring adjacent (*ortho*) to the *para*-methoxy group increases attraction about 1000 times. The overall size of the alkoxy-groups is important, and attraction rapidly decreases with increasing size; e.g. in the *ortho*-dialkoxybenzenes the LR values are: CH_3O (1.0), C_2H_5O (5), C_3H_7O (100), $(CH_3)_2CHO$ (50–100), and C_4H_9O (10,000 μg) (Metcalf et al. 1979).

LR and Hydrophobic Effects.

The lipophilicity of the side chain is important, and attractivity is increased with substituents in the 1-position of the phenyl ring having positive π values (log octanol/H_2O partition), as shown in Figure 5.4. The

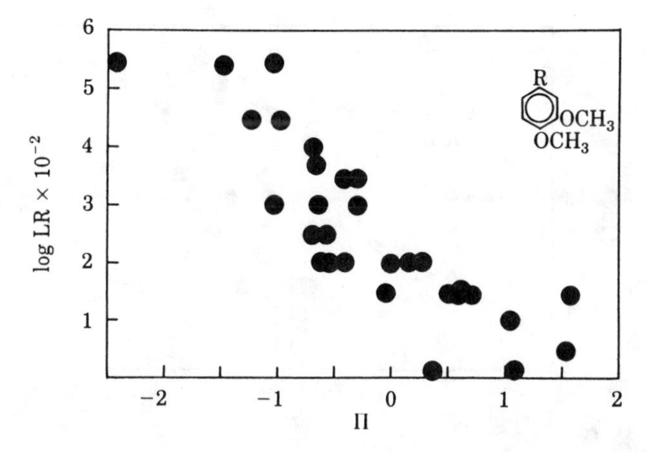

Figure 5.4. Correlation between LR (in μg) and π value for primary substituent (R) of 3,4-dimethoxybenzenes exposed to male *Dacus dorsalis*. Reprinted with permission from Metcalf et al. (1981). π = log P_R/P_H where P_H is the octanol/water partition coefficient of the unsubstituted phenyl compound and P_R is that of the phenyl compound with substitutent R. π is a measure of hydrophobic bonding.

correlation coefficient between π values for 29 side chain substituents of *o*-dimethoxybenzene was $r = -0.79$ ($P \leqq 0.01$) (Figure 5.4), demonstrating that hydrophobic interactions facilitate adsorption of the odorant molecules on the lipoprotein patch of the receptor macromolecule complementary to the $CH_2CH = CH_2$ side chain of the methyl eugenol molecule (Metcalf et al. 1981). The importance of lipophilicity is also demonstrated by comparisons of the LR values for 3,4-dimethoxybenzoic acid (1000) and that of its methyl ester (100), and of 3,4-dimethoxyphenylacetic acid (1000) and that of its methyl ester (1.0) (Metcalf et al. 1981).

LR and Electronic Effects.

The electron donating or withdrawing properties of the side chain of *o*-dimethoxybenzenes, as measured by σ values, had much smaller effects. The correlation coefficient between σ values for 23 substituents of *o*-dimethoxybenzene and LR values, was $r = 0.32$, $P = < 0.1$ (Fig 5.5). Therefore it appears that substituents donating electrons ($< \sigma$ values) produce stronger odorants than those withdrawing electrons ($+ \sigma$ values) (Metcalf et al. 1981). This suggests that increasing the electron-density around the methoxy groups perceptibly increases receptor-binding and consequently enhances attractivity.

Role of Side Chain Unsaturation.

The $C = C$ bond of the methyl eugenol molecule has evolutionary significance in receptor interaction, and the LR values for male *D. dorsalis* are 3,4-dimethoxy-1-allylbenzene (I) (LR = 0.001–0.01) > 3,4-dime-

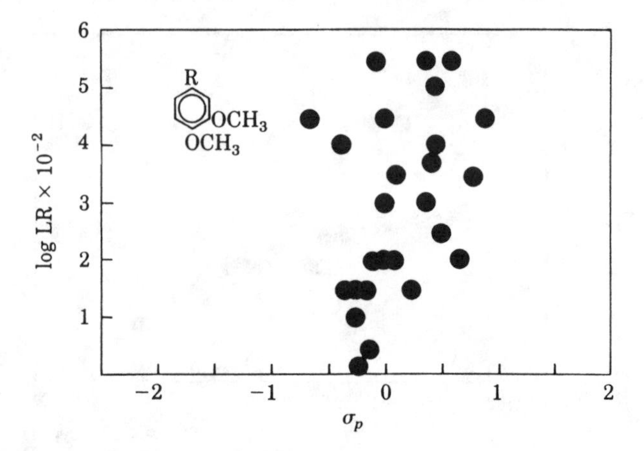

Figure 5.5. Correlation between LR (in ug) and σ value for primary substituent (R) of 3,4-dimethoxybenzenes exposed to male *D. dorsalis*. Reprinted with permission from Metcalf et al. (1981). $\sigma = \log k_R/k_H$ where k_H is a reaction constant for the unsubstituted phenyl compound and k_R is that for the phenyl compound with substituent R. Therefore σ is a measure of chemical reactivity transmitted by a substituent through an aromatic system to a reaction center.

thoxy-1-propylbenzene (II) (LR = 0.03) (Metcalf et al. 1981). Thus the male oriental fruit fly can distinguish both the presence and position of the C = C bond. Interesting attractants were produced by incorporating O-atoms in the side chains of the proper 3-atom length, e.g. 3,4-dimethoxyphenyl ethyl ether (VI) (LR = 0.03) and 3,4-dimethoxybenzyl methyl ether (V) (LR = 0.1) (Mitchell et al. 1985). These compounds are bioisosteres of methyl eugenol, and it is believed that the O-atoms, with their paired unshared electrons, produce centers of electron density that mimic that of the C = C bond of methyl eugenol (Metcalf et al. 1981). These compounds may have practical value as substitute parakairomone lures for methyl eugenol (Mitchell et al. 1985). The 3,4-dimethoxyphenylacetonitrile, with the $CH_2C \equiv N$ unsaturation, also has a surprisingly low LR of 1.0. The influence of the unsaturated side chain of the methyl eugenol molecule in providing optimal receptor contact is discernible in the male *D. dorsalis* feeding behavior. With methyl eugenol, the flies feed so compulsively that they form a compact "ball" over the treated spot, and can only be removed forcibly. Although the flies are strongly attracted to and feed on the various other 3,4-dimethoxybenzenes, as shown in Figure 5.2, feeding is much less compulsive, and the attracted flies spread much more loosely over the paper, frequently wandering about as though unsatisfied (Metcalf et al. 1981).

A composite map of key receptor site interactions between methyl eugenol and the receptor site macromolecule, and incorporating these

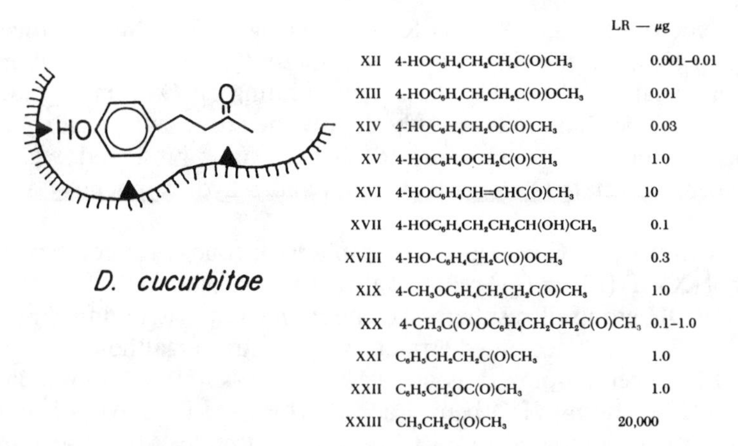

Figure 5.6. Map of suggested receptor site interactions for *D. cucurbitae* to raspberry ketone together with LR values for various analogues. Reprinted with permission from Metcalf (1990).

observations, is shown in Figure 5.2 (Metcalf 1990). Presumably all of the *Dacus* spp. responding to methyl eugenol share in these features of the receptor lipoprotein, and are descendents of an ancestral form where this kairomone dependence developed.

2. *Dacus cucurbitae* Raspberry Ketone Receptor

The ultimate sensitivity of attraction of male *D. cucurbitae* to raspberry ketone on filter paper is about 10^{-4} μg. This degree of sensitivity is approximately equal to that of *D. dorsalis* to methyl eugenol. The responsiveness of caged male *D. cucurbitae* under standardized conditions to minute residues of raspberry ketone on filter paper appears to be identical to that exhibited by male *D. dorsalis* to methyl eugenol, involving orientation, attraction, phagostimulation, and regurgitation. This response has been quantified for more than 100 compounds related to raspberry ketone (Metcalf et al. 1983, 1986). Representative compounds that are important in delineating the nature of the interaction between this plant kairomone and the receptor macromolecule are shown in Figure 5.6. The antennal kairomone receptor of the *D. cucurbitae* male is specifically complementary in size, shape, and polarity to the raspberry ketone molecule (XII), which has an LR value of 0.001–0.01 μg. LR values for *para*-kairomones were lowest with compounds approximating the raspberry ketone structure, i.e. with a *p*-OH group on the phenyl ring and a C = O group two atomic diameters from the ring, e.g. *p*-hydroxyphenyl cinnamic acid, methyl ester or methyl phloretate (XIII) (LR 0.01), and *p*-

hydroxybenzyl acetate (XIV) (LR = 0.03) (Figure 5.6). Moving the C = O group closer to the ring decreases attractivity, e.g. methyl *p*-hydroxyphenyl acetate (XVIII) (LR 0.3). Incorporating an O-atom in place of CH_2 in the side chain reduces only slightly the attractivity, as long as the proper spacing of C = O group to phenyl ring is preserved, e.g. *p*-hydroxybenzyl acetate (XIV) (LR = 0.03) and *p*-hydroxyphenoxy acetone (XV) (LR 1.0).

Reduction of the C = O group of raspberry ketone to the corresponding alcohol (XVII) (LR = 0.1) reduced attractivity substantially. Substitution of the *p*-OH group of raspberry ketone by *p*-CH_3O, as in anisyl acetone (XIX) (LR = 1.0), decreased attractivity even further, although this compound has been employed as an effective *para*-kairomone lure (Barthel et al. 1957). The *p*-CH_3O benzyl acetate (LR = 1.0) is also of reduced attractivity as compared to *p*-hydroxybenzyl acetate. Attractivity in the benzyl acetate series is abolished by incorporation of an additional *m*-CH_3O, as in 3,4-dimethoxybenzyl acetate (LR = 10,000), although this compound is reasonably attractive to *D. dorsalis* (LR = 10). These differences clearly illustrate the structural variations between the receptor macromolecules for the two fruit fly species (Metcalf et al. 1983).

Acetylation of raspberry ketone to produce cue-lure (XX), the very widely used Dacinae attractant (Beroza et al. 1960), substantially decreased attractivity in cage tests with filter paper (LR = 0.1–1.0). Cue-lure is so sensitive to hydrolysis that it is virtually impossible to measure its intrinsic attractivity because of the extreme sensitivity of male *D. cucurbitae* to the raspberry ketone formed as a hydrolysis product (Metcalf 1990, Metcalf et al. 1983). Removal of the *p*-OH group, as in phenyl-2-butanone (XXI) or benzyl acetate (XXII), greatly reduced attractivity, as did insertion of a C = C bond in the side chain (Metcalf et al. 1983).

Exploration of the relative attractivity of analogues of raspberry ketone has led to the portrayal of the interactive features of the receptor membrane, as shown in Figure 5.6. These studies led to the discovery of the highly attractive parakairomones methyl phloretate and *p*-hydroxybenzyl acetate (Metcalf et al. 1983).

Dacus cucurbitae males show slight attraction to 2-butanone or methyl ethyl ketone (Figure 5.6 XXIII). Drew (1987) has suggested that this compound, produced by bacteria on plant surfaces frequented by *D. tryoni*, is a rendezvous stimulant that brings mature male flies into feeding or oviposition sites for developing females. He proposed that 2-butanone is the primary attractive component of the cue-lure molecule (4-*p*-acetoxyphenyl)-2-butanone (XX). This idea, however, ignores the very precise receptor complimentarity with raspberry ketone, and the consequent 20×10^6 better receptor fit of raspberry ketone as compared to 2-butanone (Figure 5.6). 2-Butanone is volatilized about 1.8×10^5 times as rapidly as cue-lure (Metcalf 1990), and this very high release rate explains its initial and transient attractivity (Drew 1987). The obvious evolutionary parallel between the phenylpropanoids methyl eugenol and raspberry ketone for a large number of *Dacus* spp. cannot be explained by the attraction of 2-butanone to *D. cucurbitae* and *D. tryoni.*

D. Kairomones and Other Dacinae Species

Although at least 176 species of Dacus are attracted to cue-lure/raspberry ketone, and 58 species to methyl eugenol (Metcalf 1990), no species has been shown to respond to both lures. Moreover, the intensity of response among the various species is unlikely to be equivalent, although there are no quantitative data. The olive fly *D. oleae* is reported to be only weakly responsive to cue-lure (Orphanides et al. 1967). The response of several hundred species of Dacinae, including a number of important economic pests, (Table 5.1) has not been determined. At least 31 *Dacus* species known to be present in trapping areas were not collected by either methyl eugenol or cue-lure/raspberry ketone baited traps and are considered to be non-responsive (Drew & Hooper 1981).

Several species have shown male attraction and phagostimulation to other lures. *D. vertebratus* is strongly attracted to methyl *p*-hydroxybenzoate (Hancock 1985c). The solanaceous or Malaysian fruit fly, *D. latifrons*, inadvertently introduced into Hawaii in 1986, is not appreciably responsive to either methyl eugenol or cue-lure/raspberry ketone, but the male is attracted to and feeds upon the terpenoid α-ionol (McGovern et al. 1989). The LR value of 3 μg is much higher than those found for methyl eugenol and raspberry ketone to *D. dorsalis* and *D. cucurbitae*, respectively (unpublished data).

The discovery of new kairomone lures for unexplored species of Dacinae provides exciting opportunities for fundamental and applied research.

E. Parakairomones for *Dacus dorsalis*

Methyl eugenol, although an extraordinarily attractive and very widely used lure for the oriental fruit fly and related species (Table 5.1), has been shown to be an animal carcinogen (Miller et al. 1983). Anticipating that this finding might prejudice the future use of methyl eugenol traps for monitoring and controlling fruit flies, we investigated bioisosteres of methyl eugenol as possible replacement parakairomones (Mitchell et al. 1985). Requirements for substitute lures imposed three major restrictions

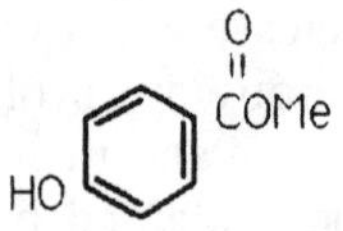

methyl p-hydroxybenzoate

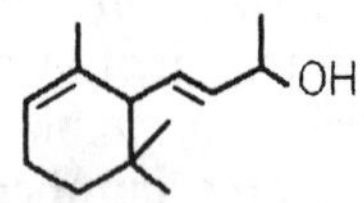

α-ionol

Table 5.9. Preferences of Male Oriental Fruit Flies for Methyl Eugenol and Parakairomones on Filter Papers Treated with 1 μg*

Attractant	Mean no. (± S.D.) flies on treated paper after (n=4). 1 min.	2 min.	5 min.	10 min.[1]
3,4-$(CH_3O)_2C_6H_3CH_2CH{=}CH_2$	8±4	12±3	16±5	17±5a
3,4-$(CH_3O)_2C_6H_3CH_2CH_2CH_3$	6±3	7±4	9±5	12±5ab
3,4-$(CH_3O)_2C_6H_3OCH_2CH_3$	5±3	7±3	9±4	13±5ab
3,4-$(CH_3O)_2C_6H_3CH_2OCH_3$	1±1	2±1	4±2	7±2b

* Data from Mitchell et al. (1985).

[1] means followed by different letters are significantly different ($P \leq 0.05$).

upon the molecular architecture of the parakairomones: (1) the candidate lure must be complementary in structure to the molecular architecture of the macromolecular antennal receptor site of *D. dorsalis*, and when adsorbed must produce receptor depolarization, (2) the candidate lure must not be susceptible to metabolic transformations in the mammalian liver to form either the 2′, 3′-epoxide or the 1′-hydroxy-derivatives of methyl eugenol. These are the putative electrophilic alkylating agents promoting liver carcinogenesis (Mitchell et al. 1985), and (3) the candidate lure must have a sufficiently high release rate to be effective as a long-range attractant for *D. dorsalis*. From extensive studies of the relationship of the chemical structure of methyl eugenol derivatives to attractivity to *D. dorsalis* males (Metcalf et al. 1981), we selected three compounds that had LR values of 0.1 μg or less (Figure 5.2, II, V, VI). None of these candidate lures had the unsaturated 1′- or 2′-propenyl side chains characterizing the most effective lures for *D. dorsalis* (Beroza & Green 1963). Therefore, *in vitro* formation of the ultimate carcinogenic moiety, 3,4-dimethoxy-1′-hydroxy-2′,3′-epoxypropane, was precluded (Mitchell et al. 1985). Laboratory preference tests in cages with 100 male *D. dorsalis* made with methyl eugenol and the three candidate parakairomones applied to filter papers at 1 μg, showed that the substitutes were all attractive in the presence of methyl eugenol (Table 5.9).

Table 5.10. Methyl Eugenol and Parakairomones as Lures for Male Oriental Fruit Flies in Jackson Traps Treated with 100 mg*

Attractant	Mean no. (± S.D.) flies caught after (n=4).				
	15 min.	30 min	60 min.	2 hr.	3 hr.
methyl eugenol	36±11	43±9	54±16	76±28	99±33a[1]
3,4-dimethoxy-propylbenzene (II)	9±3	14±7	16±7	28±5	47±10b
3,4-dimethoxyphenyl ethyl ether (VI)	6±4	7±4	14±6	23±7	29±8b

* Data from Mitchell et al. (1985).

[1] means followed by different letters are significantly different ($P \leq 0.05$).

1. Field Evaluation of Parakairomones for Dacus dorsalis

Methyl eugenol, 3,4-dimethoxy-1-propylbenzene, and 3,4-dimethoxy-1-ethoxybenzene were evaluated for attractivity to wild *D. dorsalis* in field preference tests in a grapefruit orchard at the University of Hawaii Poamoho Field Station. 100 mg of lure was used on cotton dental wicks in Jackson sticky traps. The traps were hung at head height in grapefruit trees, and were at least 30 m apart. The results, shown in Table 5.10, indicate that methyl eugenol was approximately twice as effective in attracting male *D. dorsalis* as 3,4-dimethoxy-1-propylbenzene, and about 3 times as effective as 3,4-dimethoxy-1-ethoxybenzene (Mitchell et al. 1985). It was concluded from this and other field evaluations that both of these parakairomones would be satisfactory field lures for male oriental fruit flies.

F. Parakairomones for *Dacus cucurbitae*

The incorporation of the active moiety of a kairomone lure, e.g. raspberry ketone, into a derivative such as cue-lure which serves to facilitate release rate and atmospheric transport, and which subsequently forms the kairomone through hydrolysis by atmospheric moisture, is an innovative approach to the development of parakairomones.

Laboratory preference tests. Pairs of filter papers containing equivalent doses of two different lures were exposed to 100 male *D. cucurbitae*, and the numbers of flies arrested and feeding on each paper were recorded at intervals of 1,2,5, and 10 minutes (Metcalf et al. 1983, 1986). Effects of hydrolysis were measured by comparing the attractivity of the parakairomones on a dry filter paper with that on a filter paper wetted with 0.1 ml of water (Metcalf & Mitchell 1990). The results, shown in Table 5.11,

Table 5.11. Preferenes of Male Melon Flies for Dry (D) and Wet (W) Filter Papers Treated with 0.1 μg of Aliphatic Esters of Raspberry Ketone*

Ester		Mean no. (± S.D.) flies on treated paper after (n=4). 1 min.	2 min.	5 min.	10 min.	Ratio W/D
acetyl	D	1±1	2±2	3±3	5±7	
						2.7
	W	6±8	7±7	8±8	6±3	
propanyl	D	1±1	3±4	1±1	5±5	
						8
	W	4±4	4±4	8±7	5±1	
butanyl	D	1±1	0	0.5±1	1±1	
						14
	W	8±3	8±1	7±3	6±1	
control	D	0	0	0	0	
	W	0	1±1	0	1±2	

* Data from Metcalf & Mitchell (1990).

demonstrate that moisture rapidly increased the attractivity of the acetyl, propanyl, and butanyl esters of raspberry ketone.

Effects of volatility were measured by comparing raspberry ketone and its formyl, acetyl, propanyl, butanyl, and pentanyl esters in paired laboratory preference tests with male *D. cucurbitae,* as shown in Table 5.12 (Metcalf & Mitchell 1990). These preference tests demonstrated the combined effects of release rate and hydrolysis rate in determining the responses of the male melon fly to the several parakairomones. The fruit flies were about equally responsive to raspberry ketone and the formyl and acetyl esters (Table 5.12 A,B) where the greater sensitivity (lower limit of response) to raspberry ketone and the higher release rates of the esters was essentially balanced. The less volatile propanyl and butanyl esters were not as readily perceived by the fruit flies, however (Table 5.12 C,D). In the comparisons of the several esters of raspberry ketone (Table 5.12 E,F,G,H), the formyl ester was the most attractive. These laboratory preference test data are of interest in explaining the results of the parakairomone field preference data discussed below.

1. Field Evaluation of Parakairomones for Dacus cucurbitae

Raspberry ketone and its formyl, acetyl, propanyl, butanyl, and pentanyl esters were compared in replicated field trials with a wild melon fly population at Wahiawa Public Gardens in Hawaii. These esters were applied to cotton dental wicks at equivalent 0.1 mmole doses (25 to 37 mg per lure depending upon the molecular weight), fixed in Jackson sticky traps placed 20 m apart at shoulder height. As shown by the two day accumulated trap catches (Figure 5.7) the formyl ester was approximately twice as effective in attracting flies as the acetyl ester (cue-lure) and the

Table 5.12. Preferences of Male Melon Flies for Raspberry Ketone and its Aliphatic Esters on Filter Papers Treated with 0.1 μg*

	Lure	Mean no. (± S.D.) flies on treated paper after (n=4).			
		1 m	2 m	5 m	10 m
A.	raspberry ketone	2±1	5±6	7±3	11±3
	formyl ester	4±1	9±1	16±4	17±3
B.	raspberry ketone	2±1	8±2	8±3	13±3
	acetyl ester	3±4	7±9	7±4	9±4
C.	raspberry ketone	6±5	6±1	7±6	11±3
	propanyl ester	1±1	2±1	3±3	3±4
D.	raspberry ketone	4±1	10±5	13±6	14±3
	butanyl ester	0	0	0	0
E.	formyl ester	4±2	6±3	10±2	10±1
	acetyl ester	4±3	4±2	5±2	6±2
F.	formyl ester	4±3	7±5	10±6	10±5
	propanyl ester	0	2±2	1±1	1±2
G.	formyl ester	3±2	5±2	9±4	10±4
	butanyl ester	1±1	1±1	1±1	2±3
H.	formyl ester	9±4	10±4	14±6	17±7
	pentanyl ester	1±1	1±1	1±1	0
I.	acetyl ester	1±2	2±2	4±2	3±2
	propanyl ester	1±2	2±3	4±3	4±3
J.	acetyl ester	2±2	2±2	4±1	5±2
	butanyl ester	1±1	1±1	1±1	0

* Data from Metcalf & Mitchell (1990).

catches in traps baited with the other aliphatic esters declined progressively with the increased molecular weight and consequent lowered release rate. When the mean two day catches were plotted against the release rates (Table 2.1) for the volatilization of raspberry ketone and the several esters (Figure 5.7), there was an almost perfect correlation ($r^2 = 0.993$, $P \leqq 0.0001$) between trap catch and release rate (Metcalf & Mitchell 1990). This and other field evaluations have shown that the formyl ester of raspberry ketone is consistently about twice as effective as cue-lure as a field attractant for male *D. cucurbitae.*

G. Kairomone Baits for Monitoring and Controlling Dacinae Fruit Flies

Howlett (1915) first demonstrated that methyl eugenol from citronella oil was a specific attractant to males of several species of Dacinae, including the oriental fruit fly *Dacus dorsalis.* During the outbreak of this species

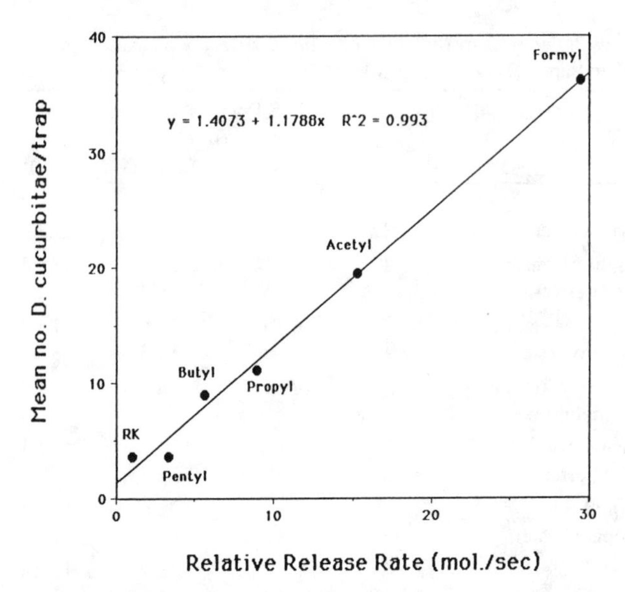

Figure 5.7. Correlation between two day trap catch of *D. cucurbitae* males and release rates of raspberry ketone and its aliphatic esters. From Metcalf & Mitchell (1990).

in Hawaii following its discovery there in May 1946, oil of citronella was evaluated by the U.S.D.A. as an attractant, and was so effective that it was used as the standard substance for monitoring oriental fruit fly infestations (Steiner 1952). Oil of citronella contains about 8% methyl eugenol, and it was soon found that the purified chemical was far superior in attracting male oriental fruit flies, for which it is also an arrestant and phagostimulant (Steiner 1952). Methyl eugenol was shown to attract male *D. dorsalis* upwind from as far as 0.5 miles, and its use for population monitoring was adapted to a variety of invaginated glass (McPhail trap) and plastic traps containing water to drown the flies. Traps containing 1 g of methyl eugenol trapped as many as 2600–7300 male *D. dorsalis* per day. The simple 8-oz (230 ml) bottle trap baited with methyl eugenol on a cotton wick became a standard monitoring device to detect incipient oriental fruit fly invasions (Steiner 1957). This has been superseded recently by the cardboard triangular Jackson sticky trap, which when baited with 1 ml of methyl eugenol is effective for about one month.

A simple box trap for area-wide control of *D. dorsalis* was made by treating the inside of 3- × 12- × 16-inch (75- × 300- × 400-mm) cardboard boxes with 0.5 g of parathion insecticide as a wettable powder, and overspraying with 2 ml of methyl eugenol (Steiner 1952). Such traps, open on one side, attracted and killed 13,000–15,000 fruit flies per trap.

1. Male Annihilation Method

The box trap with methyl eugenol and parathion was used with marked flies to show that these could be attracted from as far as 1–1.5 miles (1.6–2.4 km). In a 125-acre (50 ha) pineapple field, 45 of these box traps killed thousands of male flies and substantially reduced the male *D. dorsalis* population over an area of at least 4 square miles (10.4 km^2) (Steiner 1952). This large scale technique was termed "male annihilation". It was refined by employing cane fiber blocks 2.5 inches (62.5 mm) square and 0.37 inch (9.25 mm) thick saturated with a bait mixture of 97% methyl eugenol and 3% naled insecticide (dimethyl 1,2-dibromo-2,2-dichloroethyl phosphate), so that each fiber block contained about 23.3 g of methyl eugenol and 0.7 g of insecticide. These were dropped from aircraft at the rate of 125 per square mile (2.6 km^2) over the island of Rota in the Mariannas at about 2-week intervals for 8 months. The *D. dorsalis* population was monitored by trapping with methyl eugenol traps and declined from an average pretreatment count of 262 male *D. dorsalis* per trap to 18.4 males per trap after the first treatment and to 0.028 males per trap after the fourth treatment. No flies were caught after the seventh month of baiting, and the oriental fruit fly population was reduced by at least 99.6% (Steiner et al. 1965). This extremely efficient control effort used only 3.5 g of insecticide per acre (0.4 ha) per application, and remains a classic demonstration of the efficiency and effectiveness of kairomone lures for insect pest control.

Raspberry ketone (Willison's lure) and cue-lure have been widely used for monitoring populations of a variety of *Dacus* spp. responsive to these lures (Table 5.1), e.g. the melon fly, *D. cucurbitae* and the Queensland fruit fly *D. tryoni* (Drew 1974, Drew & Hooper 1981). Cue-lure is superior for this use because of its 15 times greater release rate as compared to raspberry ketone (Table 2.1). Surveillance monitoring with cue-lure and methyl eugenol in inexpensive Jackson traps is estimated to capture at least 90% of all Dacinae species (Hardy 1979). Surveillance for exotic Tephritidae has been routinely conducted in New Zealand fruit orchards since the mid-1970's using Jackson traps baited with methyl eugenol and cue-lure for Dacinae species and trimedlure for Ceratitinae species, and no tephritids have been captured (Somerfield 1989). Similar surveillance in the vicinity of the Los Angeles airport during 1987 revealed the inadvertent introduction of 9 species of fruit flies, including the oriental fruit fly, the melon fly, the peach fruit fly *D. zonatus*, an undescribed species of *Dacus*, and the Mediterranean fruit fly (Carey & Dowell 1989).

The male annihilation technique has been used successfully to control *D. cucurbitae* on the Island of Hawaii using fiberboard blocks 2.5 inches square and 0.5 inches thick (62 × 12 mm) treated with a mixture of 95%

cue-lure and 5% naled insecticide, so that each block contained about 23.75 g of cue-lure and 1.25 g of insecticide. The treated fiberboard blocks were tied 2–5 feet (0.6–1.5 m) above the ground on trees or stakes at the rate of 585 blocks per mi^2 (2.6 km^2) over an isolated area of the island, and were replaced with freshly treated blocks once each month for four months (Cunningham & Steiner 1972). The *D. cucurbitae* population was monitored by trapping with cue-lure bait, declining from a pretreatment count of 169 male melon flies per trap after the first treatment (96.3% decline) to 0.11 male flies per trap after the fourth treatment (99.9% decline). This level of population reduction was maintained over a period of 60 days. This very efficient control effort used only 1.2 g of insecticide per acre per treatment, and again illustrates the extremely high efficiency of kairomone baiting and the safety and specificity of this method for insect pest control.

The male annihilation technique with poisoned kairomone baits has been used successfully to eradicate *Dacus* spp. from insular populations. The use of the bait mixture of 97% methyl eugenol and 3% naled insecticide, impregnated at 24 g per cane fiber square and dropped by air at 1 to 2 per ha, eradicated *D. dorsalis* from the Island of Rota (Steiner et al. 1965). *D. tryoni* was eradicated from Easter Island using a combined treatment of 2 g each of cue-lure and malathion insecticide on 25 cm pieces of cotton string distributed at 30 per ha. This treatment was supplemented by spot-spraying with protein-malathion bait spray (Bateman et al. 1973). *D. cucurbitae* was successfully suppressed for 5 months in Okinawa by infiltrating pieces of cotton rope 5 cm long and 0.7 cm in diameter with 0.83 g of 85% cue-lure and 5% naled. Following distribution of the impregnated baits, the numbers of *D. cucurbitae* decreased to about 0.16 after the first application and to 0.01 at the conclusion of the experiment (Taniguchi et al. 1988).

D. dorsalis was effectively controlled on Lambay Island near Taiwan by the use of 4 × 4 × 8.9 cm fiber board squares impregnated with 8 ml of a 97:3 mixture of methyl eugenol and naled, applied monthly at 3–4 blocks per ha (Chiu & Chu 1988).

The male annihilation technique should be used with great caution against indigenous infestations of fruit flies because, as Hardy (1979) has emphasized, only a fraction of the 1000 species of Tephritidae are of commercial importance as pests infesting cultivars. Overenthusiastic efforts to eradicate pest species by the use of kairomone lures can lead to devastation, and even the extinction of the endemic fauna, from the Oriental, Australian, and Ethiopian Regions. Therefore, no eradication programs should be undertaken until the probable environmental impacts have been fully assessed.

IV. *RHAGOLETIS* FRUIT FLIES

Several species of *Rhagoletis* fruit flies are of considerable importance as agricultural pests of temperate and subtropical regions. These are characterized by relatively narrow host ranges, and are essentially monophagous. The most important pest species are the apple maggot, *R. pomonella* (Walsh), the cherry fruit fly *R. cingulata* (Loew), the black cherry fruit fly *R. fausta* (Osten-Sacken), the western cherry fruit fly *R. indifferens* (Curran), the European cherry fruit fly *R. cerasi* (Curran & Cresson), the blue-berry maggot *R. mendax* Curran, and the walnut husk fly *R. completa* Cresson, and the walnut husk maggot *R. suavis* (Loew) (Davidson & Lyon 1987, Carey & Dowell 1989). Information about their chemical ecology is limited, though it is apparent that host plant selection is strongly influenced by both plant kairomone and visual cues (Prokopy et al. 1987).

The apple maggot is a key pest of apples, and is very injurious in the northeastern states and Canada, and as far west as the Dakotas. This insect is very difficult to control, and together with the codling moth *Cydia pomonella* (L.) and the plum cucurculio *Conatrechalus nenuphar* (Herbst), has been responsible for the great emphasis placed on apple spray schedules involving the extensive use of insecticides. Discontinuance of the use of insecticides and acaricides in a well maintained apple orchard in New York showed that the apple maggot rendered the apple crop useless after the first year, and remained the most consistently damaging pest over a 10-year period, infesting more than 75% of the fruit (Glass & Lienk 1971).

A. Life History, Appearance, and Habits

Rhagoletis pomonella has been the most intensively studied pest of this group of Tephritidae, and will serve as a model. The adult apple maggot fly is about 5 mm long, black with white bands on the abdomen, and the wings are marked with four oblique bands. In June, the female fly lays up to 300 eggs singly under the skin of the young, green apple through her sharp piercing ovipositor. The headless white maggots tunnel through the fruit and complete their growth within a few weeks after the apple drops to the ground. The larva leaves the dropped apple and pupates in the ground, passing the winter in a brown puparium. There is a single generation per year in the northern areas and a partial second generation in the southern part of the range.

B. Kairomone Attractants for *Rhagoletis* Apple Maggot Fly

Sticky board traps were first used to monitor apple maggot flies, and to study their behavior (Buriff 1973, Maxwell 1969, Neilson 1971). These studies led to the commercial development of the rectangular, yellow, sticky apple maggot trap (Pherocon AM) (Figure 2.1). Prokopy (1968a) studied the apple maggot fly's responsiveness to visual stimuli, i.e. the size, shape, and color of sticky traps. With 30 × 40 cm rectangular board traps, both sexes preferred yellow over green, orange, red, blue, and violet; and black-white to clear. It was found that wooden spheres were more attractive than equivalent sized cubes and cylinders. There was a progressive decrease in attractivity as the size of the sphere was increased from 7.5 cm to 45 cm diameter. The color preference for the spheres was red over blue, violet, or dark orange; and black over green, light orange, yellow, white and clear. There was a progressive decrease in attraction when the spheres were dark red, but an increase when they were yellow. A suggested explanation was the association of yellow with foliage and feeding sites, together with the red spheres as oviposition sites (Prokopy 1968b). "Stickem" coated red spheres were more attractive to the flies than "Tangle-trap®" coated red spheres or "Stickem" coated red Winesap apples.

Olfactory stimulation of apple maggot flies was found to affect the capture of marked flies (Maxwell & Parsons 1968, Neilson 1971), and the distribution of the marked flies was influenced by tree variety, the presence or absence of fruit, and a preference for early and mid-season ripening varieties. Prokopy et al. (1973) found that both male and female flies responded positively to the volatile odors emanating from apples in boxes covered with cheese cloth, so that their shape was obscured. Extracts from stored, whole Red Delicious and Red Astrachan apples were assayed in a wind tunnel, significantly increasing the numbers of sexually mature males and females responding to the source (Fein et al. 1982). The crude extracts were fractionated by gas-liquid chromatography (GLC) and the behaviorally active components were found to be a series of C_8 to C_{10} esters, as shown in Table 5.13. The wind tunnel behavior and electroantennograms of the flies indicated that propyl hexanoate, $CH_3(CH_2)_5O(CH_2)_2CH_3$, was the most attractive of the volatiles, but none of the components alone were as attractive as the synthetic mixture, which was equally as attractive as the natural apple extract (Table 5.13).

In another study, more than 60 volatile esters were isolated from the apple cultivars, Royal Red Delicious, Red Astrachan, McIntosh, and Wealthy, and from hawthorne fruit, *Crategus*. 52 of these were identified chemically (Carle et al. 1987). There were many similarities between these volatiles during the July to September period of maximum activity of

Table 5.13. Attraction of Apple Maggot Flies to Apple Volatiles in Wind Tunnel*

Volatile	Flies attracted
hexyl acetate	10f
(*E*)-2-hexenyl acetate	12ef
butyl 2-methyl butanoate	18d
propyl hexanoate	36b
hexyl propanoate	30bc
butyl hexanoate	23cd
hexyl butanoate	22cd
synthetic mixture	54a
apple extract	53a
control	3

* Data from Fein et al. (1982).

the apple maggot fly. The ratios of the volatiles were variable, but the greatest similarities were found in cultivars that had similar seasonal maturity. It was concluded that apple maggot fly may be selectively tuned to the hawthorne volatiles of its wild host possibly accounting for the extreme susceptibility of the red astrachan apple, which has a very similar volatile spectrum.

Subsequent investigations of the olfactory responses of the apple maggot fly to volatile esters showed that for the acetates through hexanoates, the maximum electroantennogram amplitude was elicited with an alcohol chain length of C_9; and for heptanoates through nonanoates by alcohols with a chain length of C_{10} (Averill et al. 1988). Compared to the antennogram results, only butyl and pentyl hexanoates, propyl and butyl heptanoates, and propyl octanoate were highly attractive in wind tunnel evaluations. This demonstrated that the flies had a high degree of olfactory discrimination and also demonstrated the hazards of relying upon antennogram responses without accompanying behavioral data. It was concluded that for the maximum behavioral response the ester must (1) be straight chain, (2) be C_{10} to C_{11} in length, (3) have an acid portion of C_6 to C_8 and an alcohol portion of C_3 to C_5. Butyl hexanoate, $CH_3(CH_2)_5OC(O)(CH_2)_2\ CH_3$, found in significant amounts in the head space volatiles of apple fruits, elicited the most pronounced behavioral response from the apple maggot flies. These results suggest that *Rhagoletis pomonella* is adapted to the perception of a kairomone that is most typical of its plant hosts. This attractant has been employed in subsequent field studies with attractant baited red spheres.

C. Kairomone Lures for Monitoring and Controlling *Rhagoletis*

Stickly red wooden sphere traps for adult apple maggot flies, baited with the synthetic blend of apple volatiles, captured significantly more male and female apple maggot flies than identical unbaited spheres (Reissig et al. 1982). However, the synthetic volatile mixture did not increase the trapping efficiency of the yellow-panel (Pherocon AM®) traps during early season experiments.

In apple pest management studies in an unsprayed orchard (Reissig et al. 1984) a variety of options were investigated: (1) insecticide applications on a standard protective schedule, (2) insecticide applications based on yellow sticky panel trapping, (3) insecticide applications based on fruit maturation, and (4) trapping apple maggot flies using sticky red spheres which were non-baited, or baited with apple volatiles. None of the IPM (Integrated Pest Management) programs were completely successful because of very high apple maggot populations. The use of sticky red spheres did substantially reduce apple maggot damage, although control was not improved by the addition of the volatile kairomones. This was attributed to the small, closely spaced trees which were liberally trapped.

Apple orchard dispensers for butyl hexanoate were developed from polyethylene vials containing 4 ml of the lure (Prokopy et al. 1990). These released butyl hexanoate at about 0.5 mg per hour at 25° C, attracting apple maggot adults upwind from at least 4 m away. The rate of release was equivalent to that of 700 ripe Red Delicious apples, and the total quantity in each dispenser was sufficient to last through apple harvest. The lure dispensers were fastened to apple twigs within 10–15 cm of the 8 cm diameter red spheres coated with Tangletrap®. The combined visual-odor trapping system was evaluated as a peripheral interception system for apple maggot flies migrating into the orchard. When the traps were placed 5 m apart in perimeter trees, they were highly effective in preventing apple maggot fly immigration into the orchard interior, and in preventing fruit damage (Prokopy et al. 1990). Thus the apple maggot fly lure system could become a useful component of a new apple IPM program.

REFERENCES

Averill, A.L., W.H. Reissig and W.L. Roelofs. 1988. Specificity of olfactory responses in the tephritid fruit fly, *Rhagoletis pomonella*. Entomol. Exp. Appl. 47: 211–222.

Barthel, W.F., N. Green, I. Keiser, and L. F. Steiner. 1957. Anisyl acetone, synthetic attractant for male melon fly. Science 126: 654.

Bateman, M.A. 1972. Ecology of fruitflies. Annu. Rev. Entomol. 17: 394–418.

Bateman, M.A., V. Insungza and P. Arreta. 1973. The eradication of Queensland fruitfly from Easter Island. F.A.O. Plant Protection Bull. 21: 114.

Bauer, L., A.J. Birch and A.J. Ryan (1955). Studies in relation to biosynthesis. VI. Rheosmin. Austral. J. Chem. 8: 534–538

Beroza, M., B.H. Alexander, L.F. Steiner, W.C. Mitchell and D.H. Miyashita. 1960. New synthetic lures for the male melon fly. Science 131: 1044–1045.

Beroza, M. and N. Green. 1963. Materials tested as insect attractants. U.S. Dept. Agr. Handbk. 237, Washington, D.C.

Beroza, M., N. Green, S.I. Gertler, L.F. Steiner and D.H. Miyashita. 1961. Insect attractants. New attractants for the Mediterranean fruitfly. J. Agr. Food Chem. 9: 361–365.

Buriff, C.R. 1973. Recapture of released apple maggot flies in sticky-board traps. Environ. Entomol. 2: 757–758.

Carey, J.R. and R.V. Dowell. 1989. Exotic fruit fly pests and California agriculture. Calif. Agr. May-June: 38–40.

Carle, S.A., A.L. Averill, G.S. Rule, W.H. Reissig and W.L. Roelofs. 1987. Variation in host fruit volatiles attractive to apple maggot fly. (*Rhagoletis pomonella*). J. Chem. Ecol. 13: 795–805.

Chiu, H-t and Y-i Chu. 1988. The male annilhilation of oriental fruit fly on Lambay Island. Chin. J. Entomol. 8: 81–94.

Citrograph. 1990. What if the medfly really comes to California. 6 (1): 21–22.

Corey, E.J. and D.S. Watt. 1973. A total synthesis of (I)-α-and (I)-β-copaene and ylangene. J. Amer. Chem. Soc. 95: 2303–2311.

Cunningham. R.T. and L.F. Steiner. 1972. Field trials of cue-lure + naled on saturated fiber block board for control of the melon fly by the male annihilation technique. J. Econ. Entomol. 65: 505–509.

Davidson, R.H. and W.F. Lyon. 1987. "Insect Pests of Farm, Garden and Orchard." 8th. ed. Wiley & Sons, N.Y.

Drew, R.A.I. 1974. The responses of fruit fly species (Diptera: Tephritidae) in the South Pacific area to male attractants. J. Austral. Entomol. Soc. 13: 267–270.

Drew, R.A.I. 1975. Zoogeography of Dacini (Diptera: Tephritidae) in the South Pacific Area. Pacific Insects 16: 441–454.

Drew, R.A.I. 1987. Behavioral strategies of fruit flies of the genus *Dacus*, significant in mating and host plant relationships. Bull. Entomol. Res. 77: 73–81.

Drew, R.A.I. 1989. The tropical fruit flies (Diptera: Tephritidae: Dacinae) of the Australasian and Oceanean region. Mem. Queensland Mus. 26: 521 pp.

Drew, R.A.I., A.C. Courtice and D.S. Teakle. 1983. Bacteria as a natural source of food for adult fruit flies (Diptera: Tephritidae). Oecologia 60: 279–284.

Drew, R.A.I. and G.H.S. Hooper. 1981. The response of fruit fly species (Diptera: Tephritidae) to various attractants. J. Austral. Entomol. Soc. 20: 201–208.

Drew, R.A.I., G.H.S. Hooper and M.A. Bateman. 1978. "Economic Fruit Flies of the South Pacific Region". 137 pp., M.D. Romig, Queensland.

Drummond, R., E. Groden, and R.J. Prokopy. 1984. Comparative efficiency and optimal positioning of traps for monitoring apple maggot flies (Diptera: Tephritidae). Experimental Entomol. 13: 232–235.

Fein, B.L., W.H. Reissig, and W.L. Roelogs. 1982. Identification of apple volatiles attractive to the apple maggot *Rhagoletis pomonella*. J. Chem. Ecol. 8: 1473–1487.

Flath, R.A. and K. Ohinata. 1982. Volatile components of the orchid *Dendrobium superbum*. Rchb. f. J. Agr. Food Chem. 30: 841–842.

Fletcher, B.S. 1987. The biology of Dacine fruit flies. Annu. Rev. Entomol. 32: 115–144.

Fletcher, B.S., M.A. Bateman, N.K. Hart, and J.A. Lamberton. 1975. Identification of a fruitfly attractant in the Australian plant *Zieria smithii* as *O*-methyl eugenol. J. Econ. Entomol. 68: 815–186.

Friedrich, H. 1976. Phenylpropanoid constituents of essential oils. Lloydia 39: 1–7.

Gertler, L. I., L.F. Steiner, W.C. Mitchell, and W.C. Barthel. 1958. Esters of 6-methyl-3-cyclohexene-1-carboxylic acid as attractants for the Mediterranean fruit fly. J. Agr. Food Chem. 6: 592–594.

Glass, E.H., and S.E. Lienk. 1971. Apple insect and mite populations developing after discontinuance of insecticides: 10 year record. J. Econ. Entomol. 64: 23–26.

Guiotto, A., U. Fornasiero, and F. Baccichetti. 1972. Investigations of attractants for males of *Ceratitis capitata*. Farmaco ed. Sci. 27: 663–669.

Hagen, K.S., W.W. Allen, and R.L. Tasson. 1981. Mediterranean fruit fly: the worst is yet to come. Calif. Agr. 35 (3–4): March-April 5–7.

Hancock, D.L. 1985a. New species of African Ceratitinae. Arnoldia Zimbabwe 9 (21): 291–297.

Hancock, D.L. 1985b. New species and records of African Dacinae. Arnoldia Zimbabwe 9: 299–314.

Hancock, D. 1985c. A specific male attractant for the melon fly, *Dacus vertebratus*. Zimbabwe Sci. News 19: 118–119.

Hanson, K.R. and E.A. Havir. 1979. An introduction to the enzymology of phenylpropanoid biosynthesis, pp. 91–127 in T. Swain, J.B. Harborne, and C. van Summers, eds. Recent Adv. Phytochem. 11. Appleton-Century Crofts, N.Y.

Hardy, D.E. 1973. The fruit flies (Tephritidae - Diptera) of Thailand and bordering countries. Pacific Insects Monog. 31: 1–353.

Hardy, D.E. 1974. The fruit flies of the Phillipines (Diptera - Tephritidae). Pacific Insects Monog. 32: 1–266.

Hardy, D.E. 1979. Economic fruit flies of the South Pacific Region. Book Review. Pacific Insects 20: 429–432.

Howlett, F.M. 1912. The effect of oil of citronella on two species of *Dacus*. Entomol. Soc. London, Trans. 60: 412–418.

Howlett, F.M. 1915. Chemical reactions of fruit flies. Bull. Entomol. Res. 6: 297–305.

Itokawa, H., Y. Oshida, A. Ikuta, H. Inatomi, and T. Adachi. 1983. Phenolic plant growth inhibitors from the flowers of *Cucurbita pepo*. Phytochem. 21: 1935–1937.

Kapoor, V.C. and M.L. Agarwal. 1983. Fruit flies and their increasing host plants in India, pp. 252–257, in R. Cavalloro ed. "Fruitflies of Economic Importance." Balkema, Rotterdam, Netherlands.

Kawano, Y., W.C. Mitchell, and H. Matsumoto. 1968. Identification of the male oriental fruit fly attractant in the golden shower blossom. J. Econ. Entomol. 61: 986–988.

Keiser, I., S. Nakagawa, R.M. Kobayashi, D.L. Chambers, T. Urago, and R.E. Doolittle. 1973. Attractiveness of cue-lure and the degradation product 4-(*p*-hydroxyphenyl)-2-butanone to male melon flies in the field in Hawaii. J. Econ. Entomol. 66: 112–114.

Lewis, J.A., C.J. Moores, M.T. Fletcher, R.A.I. Drew and W. Kitching. Volatile compounds from the flowers of *Spathiphyllum cannaefolium*. Phytochemistry 27: 2755–2757.

Lin, Y-l and C-j Chen. 1984. Studies on the constituents of aerial parts of *Scutellaria rivularis* Wall. Chem. Absts. 102: 92951m.

McGovern, T.P. and R.T. Cunningham. 1988. Attraction of Mediterranean fruit flies (Diptera; Tephritidae) to analogues of selected trimedlure isomers. J. Econ. Entomol. 81: 1052–1056.

McGovern, T.P., R.T. Cunningham, and R. Flath. 1989. Latilure, a male attractant for *Dacus latifrons* the Malaysian fruit fly. Absts. 100 th. Annu. Meeting, Entomol. Soc. Amer., Dec. 10–14, San Antonio TX, paper 1327.

McGovern, R.P., J.D. Warthen, and R.T. Cunningham. 1990. Relative attraction of the Mediterranean fruit fly (Diptera: Tephritidae) to the eight isomers of trimedlure. J. Econ. Entomol. 83: 1350–1354.

Maxwell, C. W. 1969. Observations on bird tanglefoot traps as a direct method of apple maggot control. J. Econ. Entomol. 62: 945–946.

Maxwell, C.W. and E.C. Parsons. 1968. The recapture of marked apple maggot adults in several orchards from one release point. J. Econ. Entomol. 61: 1157–1159.

Metcalf, R.L. 1985. Plant kairomones and insect pest control. Bull. Ill. Nat. Hist. Survey 33: 175–198.

Metcalf, R.L. 1990. Chemical ecology of Dacinae fruit flies. (Diptera: Tephritidae). Ann. Entomol. Soc. Amer. 83: 1017–1030.

Metcalf, R.L., E.R. Metcalf, and W.C. Mitchell. 1981. Molecular parameters and olfaction in the oriental fruit fly *Dacus dorsalis.* Proc. Nat. Acad. Sci. (USA) 78: 4007–4010.

Metcalf, R.L., E.R. Metcalf, and W.C. Mitchell. 1986. Benzyl acetates as attractants for the male oriental fruit fly, *Dacus dorsalis* and the male melon fly *Dacus cucurbitae.* Proc. Nat. Acad. Sci. (USA) 83: 1549–1553.

Metcalf, R.L., E.R. Metcalf, W.C. Mitchell, and L.W.Y. Lee. 1979. Evolution of olfactory receptor in oriental fruit fly *Dacus dorsalis.* Proc. U.S. Nat. Acad. Sci. (USA). 76: 1561–1565.

Metcalf, R.L. and W.C. Mitchell. 1990. Development of new lures for the melon fly, *Dacus cucurbitae.* Report Calif. Dept. Food Agr.

Metcalf, W.C. Mitchell, T.R. Fukuto, & E. R. Metcalf. 1975. Attraction of the oriental fruit fly, *Dacus dorsalis* to methyl eugenol and related olfactory stimulants. Proc. Nat. Acad. Sci. (USA) 72: 2501–2505.

Metcalf, R.L., W.C. Mitchell, and E.R. Metcalf. 1983. Olfactory receptors in the melon fly *Dacus cucurbitae* and the oriental fruit fly *Dacus dorsalis.* Proc. Nat. Acad. Sci. (USA). 80: 3143–3147.

Miller, E.C., A.B. Swanson, D.H. Phillips, T.L. Fletcher, A. Liem, and J.A. Miller. 1983. Structure-activities studies of the carcinogenicity in the mouse and rat of some naturally occurring synthetic alkenylbenzene derivatives related to safrole and estragole. Cancer Res. 43: 1124–1134.

Mitchell, W.D. 1965. Notes and exhibitions. Proc. Hawaii. Entomol. Soc. 19: 23.

Mitchell, W.C., R.L. Metcalf, E.R. Metcalf and S. Mitchell. 1985. Candidate substitutes for methyl eugenol as attractants for the area-wide monitoring and control of the oriental fruit fly, *Dacus dorsalis* Hendel. Environ. Entomol. 14: 176–181.

Munro, H.K. 1984. A taxonomic treatise on the Dacidae (Tephritidae: Diptera) of Africa. Entomol. Mem. Dept. Agr. Rep. S. Africa No. 61, 313 pp.

Neilson, W.T.A. 1971. Dispersal studies of a natural population of apple maggot adults. J. Econ. Entomol. 64: 648–653.

Nishida, R., K.H. Tan, M. Serit, N.H. Lajis, A.M. Sukau, S. Takanishii and H. Fukami. 1988. Accumulation of phenylpropenoids in the rectal glands of males of the oriental fruit fly Dacus dorsalis. Experientia 44: 534–536.

Orphanides, R.S., C.D. Soultanopoulos and R.E. Danielidou-Phytiza. 1967. Attractive exercie sur le *Dacus oleae* pan diverses substances organique non proteinees. Ann. Inst. Phytopathol. Benaki 4: 29–38.

Prokopy, R.J. 1968a. Sticky spheres for estimating apple maggot adult abundance. J. Econ. Entomol. 61: 1082–1085.

Prokopy, R.J. 1968b. Visual responses of apple maggot flies *Ragoletis pomonella* (Diptera: Tephritidae): orchard studies. Entomol. Exp. Appl. 11: 403–422.

Prokopy, R.J., M. Aluja and T.A. Green. 1987. Dynamics of host odor and visual stimulus interactions in host finding behavior of apple maggot flies., pp. 161–166 in V. Labeyrie, G. Fabes, and D. Lachaise eds. "Insects-Plants", Junk, Dordrecht.

Prokopy, R.J., S.A. Johnson and M.T. O'Brien. 1990. Second stage integrated management of apple arthropod pests. Entomol Exp. Appl. 54: 9–19.

Prokopy, R.J., V. Moericki, and G.L. Bush. 1973. Attraction of apple maggot flies to odor of apples. Environ. Entomol. 2: 743–749.

Reissig, W.H., B.L. Fein and W.L. Roelofs. 1982. Field tests of synthetic apple volatiles as apple maggot (Diptera: Tephritidae) attractants. Environ. Entomol 11: 1294–1298.

Reissig, W.H., R.W. Weires, C.G. Forshey, W.L. Roelofs, R.C. Lamb, H.S. Aldwinckle and S.R. Alm. 1984. Management of the apple maggot. *Rhagoletis pomonella* (Walsh) (Diptera: Tephritidae) in disease-resistant dwarf and semi-dwarf apple trees. Environ. Entomol. 13: 684–690.

Ripley, L.B. and G.A. Hepburn. 1935. Olfactory attractants for male fruit flies. Entomol. Mem. Dept. Agr. S. Africa 9: 3–17.

Rohwer, G.G. 1987. An analysis of the California Medfly eradication program, 1980–82. Citrograph 72 (Sept.): C-I.

Schinz, H. and C.F. Seidel. 1961. Nachtrag zu der Arbeit Nr. 194 im Helvetica Chimica Acta 40: 1829 (1937). Helv. Chim. Acta 44: 278.

Severin, H.H.P. and H.C. Severin. 1913. A historical account on the use of kerosene to trap the Mediterranean fruit fly (Ceratitis capitata Wied.). J. Econ. Entomol. 6: 347–351.

Shah, A.H. and R.C. Patel. 1976. Role of the tulse plant *Ocimum sanctum* in control of the mango fly, *Dacus correctus* Bezzi. Current Sci. (India) 45: 313–314.

Somerfield, K.G. 1989. Establishment of fruit fly surveillance trapping in New Zealand. New Zealand Entomol. 12: 79–81

Sonnett, P.E., T.P. McGovern and R.T. Cunningham. 1984. Enantiomers of the biologically active components of the insect attractant trimedlure. Jour. Org. Chem. 49: 4639–4643.

Steiner, L.F. 1952. Methyl eugenol as an attractant for the oriental fruit fly. Jour. Econ. Entomol. 45: 241–248.

Steiner, L.F. 1957. Low cost plastic fruit fly trap. Jour. Econ. Entomol. 50: 508–509.

Steiner, L.F., W.C. Mitchell, N. Green and M. Beroza. 1958. Effect of *cis-trans* isomerism on the potency of an insect attractant. Jour. Econ. Entomol. 51: 921–922.

Steiner, L.F., W.C. Mitchell, E.J. Harris, T.T. Kozuma and M.S. Fujimoto. 1965. Oriental fruit fly eradication by male annihilation. Jour. Econ. Entomol. 58: 961–964.

Steiner, L.F., D.H. Miyashita and L.D. Christenson. 1957. Angelica seed oils in Mediterranean fruit fly lures. Jour. Econ. Entomol. 50: 505.

Tan, K.H. 1983. Responses of *Dacus* (Diptera: Tephritidae) to *Ocimum sanctum* (Linn.) extracts and different synthetic attractants in Penang, Malaysia, pp. 513–521. In R. Cavalloro, ed. "Fruitflies of Economic Importance." Balkema, Rotterdam, Netherlands.

Taniguchi, M, H. Nakamori, H. Kakinohana, and Y. Yogi. 1988. Jap. J. Appl. Entomol. Zool. 32: 126–128.

Teranishi, R., R.G. Buttery, K.E. Matsumota, D.J. Stern, R.T. Cunningham and S. Gothilf. 1987. Recent developments in the chemical attraction for Tephritid fruit flies. Chapt. 38 in G.R. Waler, ed. Amer. Chem. Soc. Sym. Ser. 330.

Warthen, J.J. and D.O. McInnis. 1989. Isolation and identification of male medfly attractive components in *Litchii chinensis* stems and *Ficus* spp. stem exudates. Jour. Chem. Ecol. 15: 1931–1946.

6

PLANT-PRODUCED SYNOMONES AND INSECT POLLINATION

I. INTRODUCTION

This area of chemical ecology is particularly glamorous because floral volatiles, like floral coloration and morphology, are highly perceptible and generally pleasing to humans. The pollination of flowers by insects is the classic coevolutionary process (Price 1984), and the fossil record indicates that this process has been proceeding for at least 225 million years since the early Triassic, where the first fossil records of flowering plants are found (Smart & Hughes 1973). The Triassic Period was also the approximate age of proliferation of early insects into modern, winged Holometabolous orders: the Coleoptera, Diptera and Hymenoptera. The almost explosive radiation of the angiosperms during the Cretaceous Period (130 myr BP) is thought to be the result of this coevolutionary association between plants and insect pollinators.

Plant odorants are of paramount importance in attracting pollinators, although color and morphology also play substantial roles (Faegri & Van der Pijl 1978, Percival 1965). It is probable that odor in insect attraction is a more primitive characteristic than that of color, as many primitive flowers that are pollinated by beetles lack color, but are conspicuous for their strong odors. The role of odor is paramount in plants pollinated by nocturnally flying insects. Kevan & Baker (1983) state that generalizations about the attractivity of odor to pollinating insects are more difficult to make than those about color because insect odor perception is more diverse than insect visual perception.

II. INSECTS AS POLLINATORS

The most important groups of insects involved in pollination ecology are the Coleoptera, Diptera, Hymenoptera, and Lepidoptera. The Coleoptera are considered to be the most primitive pollinators of the angiosperms, and were probably associated with open, bowl-shaped flowers from which the beetles fed on floral secretions, nectar and pollen (Kevan & Baker 1983). Many modern beetle species are restricted to floral diets as adults.

The Diptera are also considered to be primitive pollinators using their suctorial or lapping mouthparts to feed on both nectar and pollen. The higher Diptera (Brachycera) are generally flower feeders. The Bombyliidae or bee flies, for example, have specialized mouthparts with elongated sucking probosci that are most suitable for visiting tubular flowers. The Syrphidae, or flower flies, are the most important anthophilous Diptera. The adults specialize in visiting flowers to feed on nectar and pollen using a variety of types of mouthparts (Kevan & Baker 1983). Other families of Diptera that are important flower visitors include the Conopidae, Tachinidae, and Anthophoridae.

Most adult Lepidoptera feed on floral nectar and may exhibit extreme modifications of the proboscis. In the Sphingidae, *Xanthopan morgani praedicta*, with a proboscis 25 to 30 cm long, is the sole pollinator of the Madagascar orchid, *Anagraecum sesquipedale*, which has a corolla 25 to 30 cm deep. Moths are largely nocturnal flower visitors, while butterflies are diurnal flower visitors.

The Hymenoptera are the most important pollinators often exhibiting very specialized coevolutionary relationships. The fig wasps, *Blastophaga* (Chalcididae), for example, are crucial to the pollination of figs (Wiebes 1979). The bees (Apidae) are the most highly specialized of all insect pollinators, and have special adaptations of the mouthparts for nectar imbibation, and of the legs for pollen collection. Bees have intricate behavioral patterns for manipulating flowers (especially of the Labiateae, Leguminaceae, Orchidaceae, Scrophulariaceae, and Violaceae) and the flowers often exhibit modifications to facilitate bee pollination with hidden rewards of nectar and pollen. More than 20,000 species of bees are involved in this pollination synonomy with the angiosperms (Kevan & Baker 1983).

The honey bee *Apis mellifera* is the most important general pollinator. The major cultivars that are almost exclusively bee pollinated include alfalfa, cotton, peanut, soybean, sugar beet, citrus and deciduous fruits, and almost all vegetables. The honey bee is extraordinarily responsive to semiochemical odorants produced by flowers, and chemical ecology plays a major role in the synonomy that induces honey bees to visit a large array of blossoms in search of nectar and pollen. The 10-segmented flagellum of the worker honey bee has about 3000 *sensilla placodea* (plate organs), mostly on the eight terminal segments (Slifer & Sekhon 1961). Single cell recordings from these plate organs have defined about ten distinctive olfactory responses with little overlap (Vareschi 1971): (1) C_1-C_{14} aliphatic acids, including the queen bee sex pheromone 9-keto-*trans*-

2-decenoic acid, (2) cyclohexanoic acid; (3) C_5-C_7 acetates and formates, (4) terpene acetates, (5) terpenes such as geraniol, nerol, citronellol, and eugenol, (6) linalool and limonene, (7) cinnamic and benzyl alcohols, (8) hydrocinnamic acid, coumarin, (9) octanol, and (10) heptanone alarm pheromone. Discrimination of these groups of synomones occurs at a level above the central nervous system, but central nervous system mediation is apparently required to discriminate between individual substances of these various reaction groups, and presumably in the learning process of odor recognition (Pham-Delegue et al. 1986).

The honey bee response to volatiles can be both simple and direct or complex. For example, the writers have observed hundreds of honey bees attracted directly upwind from a hive for nearly 200 m to a 20 mg source of phenylacetaldehyde. However, Pham-Delegue et al. (1986) isolated an attractive odor fraction from the sunflower *Helianthus annuus* (Compositae) containing 27 "polar compounds" including verbenone, *trans*-carveol, ascaridole, bornyl acetate, perillyl acetate, 2,5-decadienal, eugenol, vanillin, methyl caproate, 4,5-dihydrotheaspirone, 2-tridecanone, δ-cadinol, and propriovanillone. This fraction was believed to represent the simplified semiochemical odorant of the sunflower.

II. POLLINATION OF ORCHIDS BY SOLITARY BEES

The Orchidaceae is the largest family of flowering plants, and contains an estimated 15,000 species of 450 genera. Orchids are famed among biologists for the seemingly bizarre morphology of the flowers, that has evolved over the 200 million years of coevolution with insect pollinators. Thus many of the pollination processes in the Orchidaceae are the most complex to be found among the angiosperms (Vogel 1966). The intricacies of orchid pollination by solitary bees and wasps have fascinated biologists since the observations of Cruger in 1865, which were later publicized by Charles Darwin (1884) in his classical book on orchid pollination. Despite more than a century of intensive study however, there are many unanswered questions about the behavioral and chemical ecology of orchid-bee interactions. The solitary bees that pollinate most of the orchid species are attracted to the flowers by volatile semiochemical fragrances, and in seeking to imbibe these become involved with the intricate morphology of the flowers, so that pollinaria temporarily adhere to the bee and are subsequently transferred to another female flower (Bergstrom 1978, Kullenberg 1956, Van der Pijl & Dodson 1966, Williams 1981). Although many species of orchids produce both nectar and pollen, some of the most interesting and unexplainable of these coevolutionary relationships involve orchid species whose male flowers produce no nectar. These are visited only by male solitary bees whose ostensible reward seems to be the collection of the odorants present (Williams & Whitten 1983).

A. Volatile Synomones in Pollination of *Ophrys* Orchids

Ophrys spp. orchids are pollinated by specid wasps, *Gorytes* spp., scoliid wasps, *Campsocolia* spp. and by solitary bees of the genera *Andrena, Anthophora, Chacidoma, Chlorandrena, Colletes, Eucera, Melecta, Osmia,* and *Xylocopa.* Many of the *Ophyrs* spp. exhibit an extreme degree of insect-plant coevolution, described as "deceitful pollination", in which the male insect is lured to the orchid by volatile semiochemicals and visual cues, but receives no reward because the flowers contain no nectar and pollen. Kullenberg (1956) described pollination of *Ophrys* spp. by pseudocopulation in which the male insect, attracted by the orchid volatiles, lands on the labellum of the orchid and is sexually excited by specific terpenoids to make a copulatory attack on the hairy surface of the labellum, which has a distinctive resemblance to the female insect. Labellar hairs of the orchid stimulate the male insect to perform pseudocopulatory movements which lead to the extension of the flower's pollinaria so that pollen carried by the bee from male flowers is deposited on the stigma. Bergstrom (1978) provided good illustrations of this intricate behavior.

Volatile chemicals isolated from the flowers of *Ophrys* orchids include (Bergstrom 1978):

octanol
dodecanol
tetradecanol
hexadecenol
heptanal
dodecyl acetate
methyl oleate
ethyl oleate
geranyl butyrate
C_{12} to C_{19} hydrocarbons

α-pinene
limonene
terpinene isomers
linalool
citronellol
neral
geraniol
β-cadinene
α-copaene

The principal synomone terpenoid isolated from the labella of several *Ophrys* species is δ-cadinene. This terpenoid was at least 1000 times more active in eliciting EAG responses from male *Eucera tuberculata* and *E. lonicornis* bees than the other cadinine isomers, as well as the structurally related amorphenes, muurolines, bulgarenes, copaenes, and ylangenes.

Table 6.1. Volatile Plant Kairomones from Catasetinae and Stanhopeinae Orchids Attractive to Male Euglossini Bees*

Kairomone	Mol. Wt.	Bp °C	Kairomone	Mol. Wt.	Bp °C
anisyl acetate	180		eugenol	164	254
benzaldehyde	106	179	indole	117	254
benzyl acetate	150		ipsdienol	152	
benzyl alcohol	108	205	isoelemicin	208	
benzyl benzoate	212	323	limonene	136	176
camphene	136	160	linalool	154	198
δ-3-carene	136	167	*p*-methylanisole	122	
carvone	150	226	methyl benzoate	136	
β-caryophylline	204	ca. 250	methyl cinnamate	162	
1,8-cineole	154	176	methyl *p*-methoxy-cinnamate	192	
cinnamaldehyde	132	246	methyl phenyl acetate	150	
cinnamyl acetate	176	145	methyl salicylate	152	223
cinnamyl alcohol	134	250	myrcene	136	
p-cresol	108	202	nerol	154	224
p-cymene	134	177	β-ocimene	136	176
dihydrocarvone	152		α-phellandrene	136	171
p-dimethoxybenzene	138	213	2-phenethyl acetate	164	
methyl eugenol	178	254	2-phenethanol	122	218
α-*p*-dimethylstyrene	132		3-phenpropyl acetate	178	
			α-pinene	136	156
			terpen-4-ol	154	220

* Data from Williams and Whitten (1983).

The EAG response to δ-cadinine paralleled the EAG pattern from labellar extracts of the *Ophrys* species known to be highly attractive to *Eucera* males (Priesner 1973).

Bergstrom (1978) suggested the following role for volatile semiochemicals in the *Ophrys*-pollinator relationship. There are three regions of dispersal of the volatiles from the orchid blossom: (1) long distances of up to 500 m, where visual cues are impossible and volatile semiochemicals diffuse downwind, (2) the neighborhood of the flower from 1 to 100 cm where volatile odorants are transported by convection and by air currents, and where visual stimulation becomes effective, and (3) close to and in visual contact with the orchid, where odor diffusion, microturbulence, and chemical and visual contact are operative. Highly volatile semiochemicals of low molecular weight (Table 6.1) are thought to be responsible for the long range attraction of the pollinators, and sesquiterpenes are probably the releasers of the pseudocopulatory behavior. The association pattern of pollination, in which specific bee species pollinate specific orchid species, can be explained by the stimulation of the bee by semiochemicals such as δ-cadinene (Bergstrom 1978, Priesner 1973).

B. Volatile Synomones in Pollination of Orchids by Euglossini Bees

More than 650 species of neotropical orchids are pollinated solely by male Euglossini bees (Ackerman 1983). At least 100 spp. of *Euglossa*, 13 species of *Eulaema*, and 52 species of *Eufriesia* are free living species that forage widely in flowers for nectar and pollen. These bees have a peculiar relationship with the Orchidaceae known as "euglossini syndrome" (Williams 1981). The orchid spp. pollinated by Euglossini are highly fragrant, but lack nectar and have no available pollen food. The male bees that visit the flowers are attracted solely by volatile semiochemicals (Table 6.1). Pollination of the orchids by the male bees is effected through an elaborate behavioral sequence: (1) attraction to the fragrance, (2) alighting on the flower, (3) moving to the base of the labellum to contact maximum fragrance, (4) brushing the surface of the orchid labellum with tarsal brushes of the front legs, (5) launching into the air, (6) transferring the collected fragrances to the hind tibia, and (7) returning to the flower (Dodson et al 1969). In *Eulaema cingulata* the foretarsi contain mop-like collection brushes for collecting the orchid fragrances. These monoterpenes and aromatics (Table 6.1) are transferred to hind tibial organs that have deeply invaginated and hair-filled cavities (Whitten et al. 1989). This bee has large cephalic glands which secrete long chain aliphatics such as dodecyl, tetradecyl, and hexadecyl acetates, and C_{25}-C_{33} *n*-alkanes that the bee mixes with the collected fragrances before they are transferred to the hind tibial baskets. Whitten et al. (1989) suggest that the cephalic gland secretions serve to retard evaporation of the collected orchid fragrances.

Explanations for the behavior of the Euglossini bees attracted to the orchid fragrances remain enigmatic (Ackerman 1983, Dodson et al. 1969, Jansen et al. 1982, Williams 1983). The collected floral fragrances clearly have significance in lek formation with other males to then attract females (Williams 1983) or in attracting females for mating. Whitten et al. (1989) hypothesize that the male bees collect the fragrances in large quantities in the tibial baskets for facilitating courtship with the female bees, so that the "tibial bouquet" might serve as an indicator of male fitness. Williams (1981), however, concludes that "a satisfactory answer has not yet been provided to the question of why euglossini bees visit orchid flowers."

Dodson and co-workers (Dodson 1970, Dodson et al 1969, Dodson & Hills 1966, Hills et al. 1968, Williams & Dodson 1972) pioneered in the demonstration that volatile orchid flower fragrances were responsible for attracting Euglossini bees to orchid flowers. These studies have been summarized by Williams & Whitten (1983) with the identification of at least 40 low molecular weight terpenoids and aromatics, isolated from neotropical orchid blossoms, that are attractive to the various species of

Euglossini bees. The synomones, listed in Table 6.1, are essentially the same plant volatiles that are known to be kairomones for phytophagous insects (Table 1.1), illustrating both the economy of nature's chemical factories and the importance of suitable chemical and physical properties for volatile semiochemicals. Wierman (1970) has pointed out that high concentrations of phenylpropanoids are produced during pollen maturation in flowers.

There was strong attraction by specific synomones where purified volatiles, identified from orchid blossoms, were placed on blotters in areas inhabited by Euglossini, as demonstrated by the performance of the appropriate behavioral sequence. Cineole attracted 433 bees of 27 species, eugenol 300 bees of 7 species, and benzyl acetate 36 bees of 4 species (Williams 1981). Of the approximately 60 species of Euglossinae in Panama, cineole attracted 35, methyl salicylate 11, and benzyl acetate 6 (Williams 1983). The responses demonstrated the highly specific nature of receptor responses to specific chemical structures of the orchid volatiles. For example, eugenol attracted 300 bees of 7 species, but only a single bee was attracted by eugenol methyl ether (methyl eugenol). Methyl salicylate attracted 11 species, while methyl benzoate attracted only two species, and β-ionone attracted 5 species but its stereoisomer α-ionone attracted only one species (Dodson 1970). An artificial fragrance of cineole and benzyl acetate attracted only 8 species of Euglossini, as compared to 27 species attracted to cineole and 8 to benzyl acetate when used alone. When α-pinene was added to the fragrance mixture, attraction was even more selective, attracting only two species. This mixture was modeled on the fragrance of *Stanhopea tricornis* and one of the two species attracted, *Eulaema meriana*, is a known pollinator of this orchid (Williams 1983).

These investigations demonstrate the major role of orchid blossom fragrances in the coevolution of orchids and bees (Ackerman 1983). Olfactory cues provide a very efficient means of attracting pollinators from long distances, and the production of low molecular weight fragrances by orchid plants is decisive for the pollination of these rare and widely dispersed plants as well as to the long distance flow of pollen gametes (Williams 1981). Species-specific orchid fragrances that selectively attract species-specific pollinators are thought to be of major importance as isolating mechanisms between closely related sympatric species of orchids (Williams 1981).

REFERENCES

Ackerman, J.D. 1983. Specificity and mutual dependence of the orchid-euglossine bee interaction. Biol. J. Linn. Soc. 20: 301–314.

Bergstrom, G. 1978. Role of volatile chemicals in Ophrys-pollinator interactions. Chapt. 8

in J.B. Harborne ed. "Biochemical Aspects of Plant and Animal Coevolution.", Academic Press, N.Y.

Darwin, C. 1984. "The Various Contrivances by which Orchids are Fertilized by Insects." 2nd ed. D. Appleton, N.Y.

Dodson, C.H. 1970. The role of chemical attractants in orchid pollination, pp. 83–107 in K.E. Chambers, ed. "Biochemical Coevolution." Oregon State University Press, Corvallis.

Dodson, C.H., R.E. Dressler, H.G. Hills, R.M. Adams and N.H. Williams. 1969. Biologically active compounds in orchid fragrances. Science 164: 1243–1249.

Dodson, C.H. and H.G. Hills. 1966. Gas chromatography of orchid fragrances. Amer. Orchid Soc. Bull. 35: 720–725.

Faegri, K. and L. Van der Pijl. 1978. "The Principles of Pollination Ecology.", 3rd ed. Pergamon Press, N.Y.

Hills, H.G., N.H. Williams and C.H. Dodson. 1968. Identification of some orchid fragrance components. Amer. Orchid Soc. Bull. 37: 967–971.

Jansen, D.H., P.J. DeVries, M.L. Higgins and L.S. Kinsey. 1982. Seasonal and site variations in Costa Rican euglossine bees at chemical baits in lowland, deciduous, and evergreen forests. Ecology 63: 66–74.

Kevan, P.G. and H.G. Baker. 1983. Insects as flower visitors and pollinators. Annu. Rev. Entomol. 28: 407–453.

Kullenberg, B. 1956. Field experiments with chemical sexual attractants on aculeate Hymenopteran males. I. Zool. Bidrag Uppsala. 31: 253–352.

Kullenberg, B. and G. Bergstrom. 1974. The pollination of *Ophrys* orchids, pp. 253–258 in Bendz, G. and J. Santesson, eds. "Chemistry in Botanical Classification.", Nobel Symposium 25. Nobel Foundation, Stockholm, Sweden.

Percival, M.S. 1965. "Floral Biology." Pergamon Press, N.Y.

Pham-Delegue, M.H., C. Masson, P. Etievant and M. Azar. 1986. Selective olfactory choices of the honeybee among sunflower aromas. J. Chem. Ecol. 12: 781–793.

Price, P.W. 1984. "Insect Ecology", 2nd ed. John Wiley & Sons, N.Y.

Priesner, E. 1973. Reaktionen von Riechrezeptoren männlicher Solitarbienen (Hymenoptera; Apidae) auf Inhaltsstoffe von Ophrys-Bluten. Zoon. Suppl. 1: 43–55.

Slifer, E.J. and S.S. Sekhon. 1961. Fine structure of the sense organs in the antennal flagellum of the honeybee *Apis mellifera.* J. Morphology 109: 351–362.

Smart, J. and N.F. Hughes. 1973. The insect and progressive paleological integration, pp. 143–155, in H.F. van Emden ed. "Insect/Plant Relationships." Roy. Entomol. Soc. London Symp. No. 6.

Van der Pijl, L. and C.H. Dodson. 1966. "Orchid Flowers: Their Pollination and Evolution." University Miami Press, Coral Gables, Fla.

Vareschi, E. 1971. Duftunterscheidung bei den Honigbiene-Einzelzell-Ableitungen und Verhaltens-reaktion. Jour. Comp. Physiol. 75: 143–173.

Vogel, S. 1966. Scent organs of orchid flowers and their relation to insect pollination, pp. 253–259 in DeGarmo, L.R. ed. Proc. 5th. World Orchid Conference. World Orchid Conference, Long Beach, Cal.

Whitten, W.M., A.M. Young and N.H. Williams. 1989. Function of glandular secretions in fragrance collection by male Euglossine bees (Apidae: Euglossini). J. Chem. Ecol. 15: 1285–1296.

Wiebes, J.T. 1979. Coevolution of figs and their insect pollinators. Annu. Rev. Ecol. Syst. 10: 1–12.

Wierman, R. 1970. Die Synthese von Phenylpropanen während der Pollenwicklung. Planta 95: 133–145.

Williams, N.H. 1981. The biology of orchids and Euglossini bees, Chapt. 4 in J. Arditti ed. "Orchid Biology, Reviews and Perspectives II." Cornell University Press, Ithaca, N.Y.

Williams, N.H. 1983. Floral fragrances as cues in animal behavior, pp. 50–71 in G.E. Jones and R.J. Little eds. "Handbook of Experimental Pollination Biology." Scientific and Academic Editors, New York.

Williams, N.H. and C.H. Dodson. 1972. Selective attraction of male euglossine bees to orchid floral fragrances and its importance in long distance pollen flow. Evolution 26: 84–95.

Williams, N.H. and W.M. Whitten. 1983. Orchid floral fragrances and male Euglossine bees: methods and advances in the last sesquidecade. Biol. Bull. (Woods Hole) 163: 355–395.

INDEX